D1237092

THE UPLAND SOUTH

THE UPLAND SOUTH

The Making of an American Folk Region and Landscape

Terry G. Jordan-Bychkov

Santa Fe, New Mexico, and
Harrisonburg, Virginia

in association with
the University of Virginia Press

PUBLISHER'S NOTES:
This book was brought to publication in an edition of 1,500 hardcover copies with the generous financial assistance of the Webb Chair Endowment of the University of Texas, Austin, for which the publisher is most grateful. For more information about the Center for American Places and the publication of *The Upland South*, see page 122.

Published 2003. First edition.
Printed in Canada on acid-free paper through Global Ink, Inc.
All photographs, drawings, and maps are by the author, unless otherwise noted.

Center for American Places, Inc.
P.O. Box 23225
Santa Fe, NM 87501, U.S.A.
www.americanplaces.org
877-608-5732

Distributed by the University of Virginia Press
P.O. Box 400318
Charlottesville, VA 22904, U.S.A.
www.upress.virginia.edu
Book orders: 800-831-3406
Information: 434-924-3468

9 8 7 6 5 4 3 2 1

Library of Congress Cataloging-in-Publication Data

Jordan-Bychkov, Terry G., 1938-
The Upland South: the making of an American folk region and landscape / Terry G. Jordan-Bychkov.
p. cm.
Includes bibliographical references (p.) and index.
ISBN 1-930066-08-2 (alk. paper)
1. Southern States—History. 2. Mountain life—Southern states—History. 3. Southern States—Social life and customs. 4. Folklore—Southern States. 5. Southern States—History, Local. 6. Human geography—Southern States. 7. Landscape—Social aspects—Southern States—History. 8. Vernacular architecture—Southern States. 9. Mountain people—Southern States—History. 10. Mountain people—Southern States—Social life and customs. I. Title.

F209.J67 2003
975'.00943—dc21 2003041048

ISBN 1-930066-08-2

FRONTISPIECE: *Twin Falls State Park, Wyoming County West Virginia*. Photo by the author, 1994.

To Bella, the love of my life,
my intellectual soul mate and best friend,
my "Kiska" who brought "Bemby"
comfort, solace, and a reason to live
in the most difficult of times for both of us
and the greatest of joy in our good times.
Always remember what we shared,
good and bad alike.

"Planting is best done
in the fruitful signs of . . . Cancer."

(From Wigginton, *Foxfire Book*, Vol. 1, p. 217)

Contents

List of Figures

Figure

Preface and Acknowledgments

THIS SMALL VOLUME is a distillation of four decades of intermittent field research in the Upland South. Perhaps my fascination with this distinctive region derived mainly from its role in my ancestry and familial heritage, from North Carolina and Tennessee to Texas, for I first encountered the Upland South in genealogical research. Soon, however, I became captivated by its natural beauty and rich, archaic cultural landscape. From there it was one short step to wondering how this place and its people—my people—came to be.

In this ongoing if often interrupted inquiry, I met and was assisted by many persons, from the "real" people on the land to fellow academicians, hedonists like me drawn to the investigation of this way of life. Among the real people I would name are T. E. Mason and his wife Mary Autry Mason of the East Cross Timbers region in North Texas; and A. C. Tiller, my maternal grandfather, of East Texas. I did not record the names of the almost countless anonymous farmers who let me look inside their barns and nose around their log houses; and the nameless graveyard workers who paused, leaned on their hoes, and answered questions that must have seemed silly.

The academicians and independent scholars who shared their ideas and even their field data with me include Richard Pillsbury at Georgia State; John Rehder at Tennessee; Edward Price at Oregon; Anita Pitchford Walker; the late Wayne Price at Illinois; Don Huebner of Austin (who showed me a four-crib barn right under my nose that I had failed to find in forty years); Lynn Morrow and James Denny of Missouri, who have amassed a huge amount of data on the Ozarks; Gene Wilhelm, who knows the Blue Ridge better than anyone else; John Morgan at Emory & Henry; Goodloe Stuck of Shreveport; the late Fred Kniffen at Louisiana State; Don Ball of Louisville; Catherine Bishir of Raleigh; Richard Finch at Tennessee Tech; the late Edwin Foscue at Southern Methodist, who first drew my attention to the southern Appalachians; Marshall Gettys, who shared his Oklahoma discoveries; John Milbauer of Tahlequah in the Cherokee country; Richard Hulan, the first person to understand the culture history of the dogtrot house; Greg Jeane and Ruth Little, fellow graveyard trampers; Merilyn Osterlund; Mary Ruth Winchell; Douglas Helm; Charles L. Sullivan; the late Alice Andrews; Don Brown; J. Roderick Moore; Pat Irwin; Charlotte Daly; and Joe de Rose. Professor Ary J. Lamme III at Florida also provided particularly valuable suggestions for improving the book, as befits a truly kindred spirit.

But my greatest debts are owed to professors John B. Rehder of the University of Tennessee at Knoxville and Gerald Smith of the University of the South in Sewanee, Tennessee. John not only provided valuable suggestions for improving the manuscript, but also shared his knowledge and field data concerning East Tennessee. Jerry did the same for a region centered on southern Middle Tennessee. Both generously took the time to be my host and accompany me into the field. They are true scholars and gentlemen, and good friends in the bargain.

Joanne Sanders of the University of Texas Geography staff typed the manuscript with skill and speed, and Joy Adams, my Webb Fellow, masterfully created the computer cartography. In these two, I was privileged to work with skilled professionals.

I also owe thanks to my highly competent doctors—Robert Frachtman, Brant Victor, and, above all, Thomas Tucker, whose skills created the "chemotherapeutic" remission that allowed this book to be written. Yes, they did it the old-fashioned way—by "slash, burn, and poison"—but then cancer treatment has not changed much over the decades. They never promised me anything, declaring my illness terminal, inoperable (a condition ascertainable only through major surgery!), and incurable. But they delivered a year's remission (and more) during which I returned to a normal life.

During the illness many people gave me love, moral support, encouragement, and "good energy." Above all my dear wife Bella did this for me, her Bemby! So did my children—Tina, Sonya, and Eric—who forgave my several shortcomings and rallied around me with love. Granddaughters Madeleine, Anna Belle, and Olivia also lent their precious support.

Cousins Linward and Beverly Shivers, Betty Laird, Steph and John Mood, the late and precious Mary Lynn Weir, Lois and Bill Koock, Nelson Durst, Sylvia Griffin, and Ethyl and Harold Byrn touched me with their love, assistance, and encouragement. Brother-in-law Vova Bychkov of Moscow inspired me by example, calls, and letters. Friends Steve Hall, Greg and Sookie Knapp, Bob and Joyce Holz, Milan Reban, Andy Schoolmaster, Kirby and Teresa Smith, Joanne and Ray Sanders, Dee Dee Miller, John Cotter, Jean Andrews, "granddaughter" Shannon Crum, Vic Mote, Paul Hudson, Bill and Diane Doolittle, Judy Dykes-Hoffmann, and Sakena Sounny-Slitine all helped keep me going.

And the indomitable Celtic spirit of my departed mother Vera told me to "keep on coming," as she had done. Celts, both real and subliminal, always lose in the end, but we leave our persecutors bruised and bloody.

All stupid mistakes I may have made along the way and in this book are my own fault. Hey, nobody's perfect!

Austin, Texas, Thanksgiving, 2002

THE UPLAND SOUTH

CHAPTER I

A Highland Folk

A CULTURAL AND PHYSICAL FAULT LINE divides the American South. We have always known, intuitively if not intellectually, that the South bears this dual character, and perhaps no more misleading label has ever been coined than "the solid South."[1]

The southern dichotomy has found expression in such paired terms as upland and lowland, mountains and coastal plain, Appalachia and Deep South, or yeoman and planter. Cotton belonged to only half of the South and not all states with legalized slavery seceded from the Union. Terms such as "Up Country" and "Hill Country" long ago entered the vernacular speech and found a place on mental maps in states such as South Carolina and Texas, whose borders include parts of both Souths.

A Way of Life

The Upland South—a mountainous, hilly, and rolling land, a land of misty ridges, cozy coves, and hidden hollows—long housed a distinctive way of life. Today, this highland culture barely survives other than in a significantly altered type. Formerly, sheltered in places by montane isolation, it found a special character in its people—old-stock whites of colonial and largely British ancestry—who practiced a mixture of hardscrabble farming, free-range hill herding, hunting, gathering, and fishing. These people spoke a unique nasal dialect (Figure1.1); perpetuated diverse traditional handicrafts; passed on a rich, archaic folklore; adhered to an emotional, vigorous Calvinism splintered into myriad, often locally independent churches and sects meeting in plain board chapels (Figure 1.2); produced "moonshine" corn whiskey; spawned a

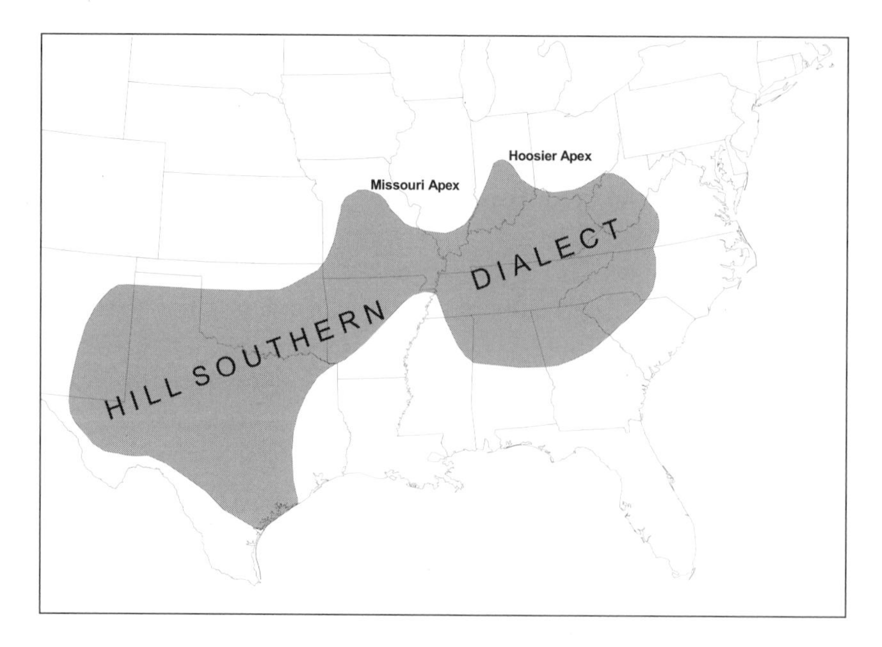

FIGURE 1.1 *The "Hill Southern" dialect region, based upon a generalization from vocabulary and pronunciation isoglosses. West of the Appalachians, the level of generalization required to draw a single line as the border becomes perhaps excessive. (Sources: Carver 1986, Kurath 1949, Pederson et al. 1986.)*

mournful "country" music; and buried their dead in distinctive graveyards that reflected a particular deathlore. For healing they relied upon an herbal-based folk medicine. Clan-based feuding, endemic to the Upland South, brought the region some of its earliest notoriety. Elements of an antecedent, frontier folk culture of the forest survived in abundance and lay at the root of many distinctive upland southern traits.[2]

FIGURE 1.2 *A typical, unadorned upland southern board chapel is the Eagle Springs Baptist Church in rural Coryell County in Central Texas. (Photo by the author, 1988.)*

Enough remnants of this highland way of life survive, especially in speechways and the commercialized versions of country music, to lend the modern Upland South a special character even today.[3] One need only read the abundant hand-painted signs along the highways to discover vestiges of the upland southern way-of-life. "Gospel singing and hot dogs" (Saluda County, South Carolina); "Falling Rock Gospel Tabernacle: Keep Looking Up" (Kanawha County, West Virginia); "Stay Out—Hogs Bite" (Lewis County, West Virginia); "Jesus loves you, drive safely, double fries" (Telfair County, Georgia); "For sale billy goat" (Overton County, Tennessee); and "Narrow Path Free Will Baptist Church" (Wayne County, West Virginia) provide representative examples.

America at large has long been fascinated by the Upland South. One need but think of the phenomenal success of the movie and soundtack of "O'Brother, Where Art Thou?" and the enduring popular parodies such as the "Beverly Hillbillies," "Dukes of Hazzard," "Mayberry R. F. D.," and "L'il Abner" in Dogpatch, of the superhuman exploits of Davy Crockett or Sergeant York, or of the touristic popularity of Branson, Missouri, or Gatlinburg, Tennessee.[4] Perhaps the nonconformity of the Upland South—its stubborn unwillingness or inability to share the dream and success of America—explains popular fascination with the region. Or perhaps its mournful music, broadcast abundantly to an entire nation, comes closer to describing the true nature of life in America for most people.[5]

Whatever the reason for America's fascination with the Upland South, scholars from a wide range of academic disciplines have, for over a century, shared it.[6] The hill folk of southern Appalachia, the Ozarks, the Shawnee Hills of southern Illinois, the Ouachita Mountains, and the hills of central Texas have been visited by anthropologists, folklorists, economists, linguists, and others almost as frequently as have certain American Indian tribes.[7] Geographers, too, have been part of the invading horde, from the time of Ellen Churchill Semple more than a hundred years ago.[8]

Boundaries

A wide variety of scholars, then, long ago concluded that a distinctive people and way of life once existed in the interior South, vivid traces of which survive into the twenty-first century. But we have by no means achieved a proper understanding of the character, origins, and geographical distribution of upland southern culture. Indeed, we cannot even agree what to call it. Linguists initially labeled its distinctive dialect "South Midland," a term accepted by a few others, then shifted their preference to "Hill Southern." Other candidates include

"Upper South," "Mountain South," "Appalachia," and, of course, "Upland South."[9] I have chosen the latter term because I find it properly inclusive, widely used, and aesthetically pleasing.

Most of all, those who have written about the Upland South disagree concerning its boundaries. Geographers, the experts most attentive to spatial distributions, have over the decades drawn widely differing lines on their maps to delimit all or part of the Upland South (Figure 1.3).[10] This should not surprise us, since cultural boundaries are generally rather fuzzy. Indeed, the search for precise borders represents a fool's errand. But these maps still instruct us, representing as they do the collective observations and intuition of several generations of talented scholars. Moreover, by shading an area mutually agreed upon by nearly all as belonging to the Upland South, we can reach some sort of consensus. The resultant region stretches from the Fall Line, at the inner margin of the Atlantic coastal plain, through southern Appalachia, across the Ohio River into southern Indiana and Illinois, where it is very nearly severed in two at the "isthmus" of the Shawnee Hills, and then southwestward through the Ozarks, Ouachitas, and the Hill Country of Texas (Figures 1.3 and 1.4).[11]

We can also learn from geographers where the Upland South is *not*. Other well-identified cultural regions line the blurred perimeter of the Upland South. Geographers have written of the Corn Belt, Dairy Belt, Cotton Belt, and Wheat Belt, all of which display cultural traits alien to the region under study.[12] Nor does the livestock ranching culture-complex of the American West belong in the Upland South. Likewise excluded are the adjacent culture regions labeled the Midwest, Pennsylvanian, Tejano Homeland, Acadiana, and Lowland South.[13] Here, again, no two geographers would agree on the precise borders of all these regions, and to draw their various lines on a map would only confuse the cartographic issue further. We must be content to know that the borders of the Upland South cannot be presented with precision and that no two traits of the upland southern culture display the same geographic distribution. We would do best to think of a core and a periphery. Henry Glassie perhaps put it best when he said "regions have fuzzy, syncretistic borders, and any attempt to define them on a map is a process of constant compromise."[14]

Whose Child?

We know less than we should about the origin and evolution of the Upland South. Because by almost every spatial definition it lacks a seacoast and, in particular, possesses no footholds in the Atlantic coastal plain, we have always assumed that the Upland South is the child of either one or another antecedent seaboard colonial cultural nucleus or of several.

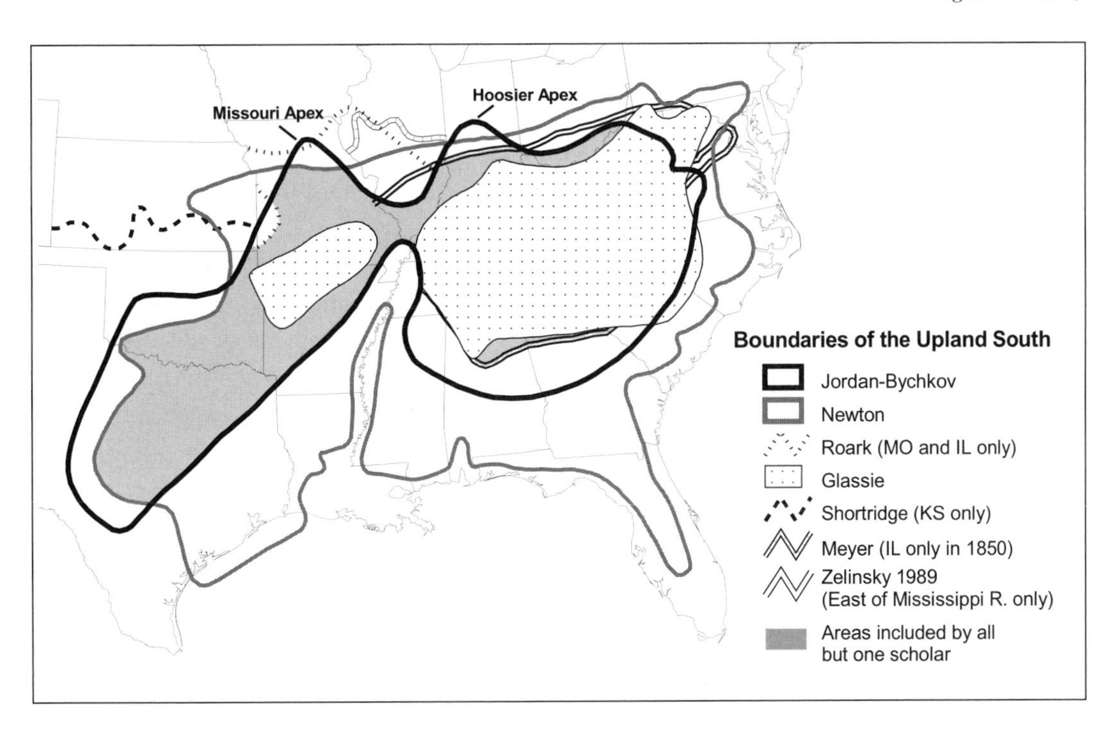

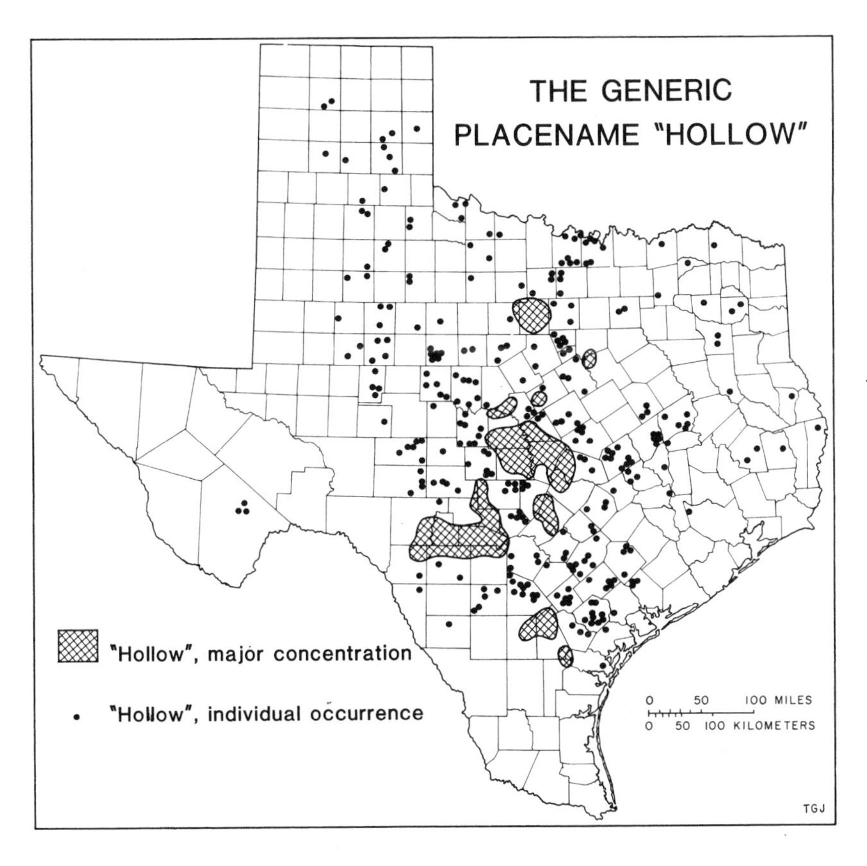

(top)
FIGURE 1.3 *Various proposed borders for the Upland South. A wide range of disagreement exists, but a core area of consensus can be seen. (Sources: Glassie 1968a, 39; Meyer 2000, 165; Zelinsky 1989, 154; Roark 1990, 16; Shortridge 1995, 187, 193; Jordan-Bychkov 1998, 6.)*

(bottom)
FIGURE 1.4 *Distribution of the generic toponym "hollow" in Texas. Meaning an elongated, flat-bottomed valley, "hollow"—pronounced "holler"—is one good index of the presence both of upland southerners and their preferred habitat. (Source: redrawn from Jordan 1970, 419.)*

Most commonly, the Upland South is presented as the child of the lower Delaware River valley implantation of Euroamericans—in short, as part of "Pennsylvania Extended" and merely the southern half of the huge Pennsylvania-derived Midland culture.[15] As noted earlier, some scholars refer to the Upland South as the "South Midland."[16] And, without question, Pennsylvanian influences exist in almost every aspect of upland southern life.

Others feel that the main thrust of Pennsylvanian influence flowed due west, into the lower Midwest—the Corn Belt. They often point to the English-dominated Chesapeake Tidewater district—the nucleus of Virginian culture—as a second important source of upland southern culture. If at least some Pennsylvanian influence moved southwestward, deflected by the Appalachians along the routes of the Great Valley, Piedmont, and Fall Line Road to reach the Upland South, the Virginian Tidewater element migrated at right angles to this flow, up the James, Rappahannock, and Potomac river valleys or by other parallel routes, and crossed through the Blue Ridge to enter the Appalachians.[17] Another contingent of lowlanders, smaller than the Virginian one, moved up from the South Carolinian hearth, entering the Piedmont and passing westward through Saluda Gap to enter the mountains by way of the Asheville Basin. These Carolinians introduced a free-range "cowpen" system of cattle and hog raising.[18]

In this view, then, the Upland South resulted from a blending of colonial Pennsylvanian, Virginian, and Carolinian subcultures. Geographer Doug Meyer perhaps said it best, pointing to the fusion of low country southern and Pennsylvanian peoples that "contributed to the creation of an Upland Southern cultural complex."[19] The regional genealogy is getting complicated.

As a result, we need to chart four levels of cultural hearths of the Upland South—not only these early colonial coastal sources, but also secondary inland centers of cultural mixing, a tertiary area of cultural ferment, and a quaternary area of final coalescence (Figure 1.5). The secondary hearths, like the primary ones, were multiple. The Upland South did not initially evolve in one highland nucleus, but rather, I feel, in at least three—the Shenandoah Valley of Virginia, a section of the Great Valley of the Appalachians; the Piedmont of northern North Carolina; and the South Carolina Up Country.[20] The influence of these secondary hearths proved long-lived. For example, as late as the twentieth century, the North Carolina Piedmont spawned the Bluegrass style of country and western music.[21]

Migration from the secondary centers, in turn, created a tertiary hearth, where the Upland South achieved its nearly final form. This lay in the Watauga country, astride the North Carolina-Tennessee border, with a focus in the northern part of the Valley of East Tennessee (Figure 1.5). The formative decade in Watauga seems to have been the

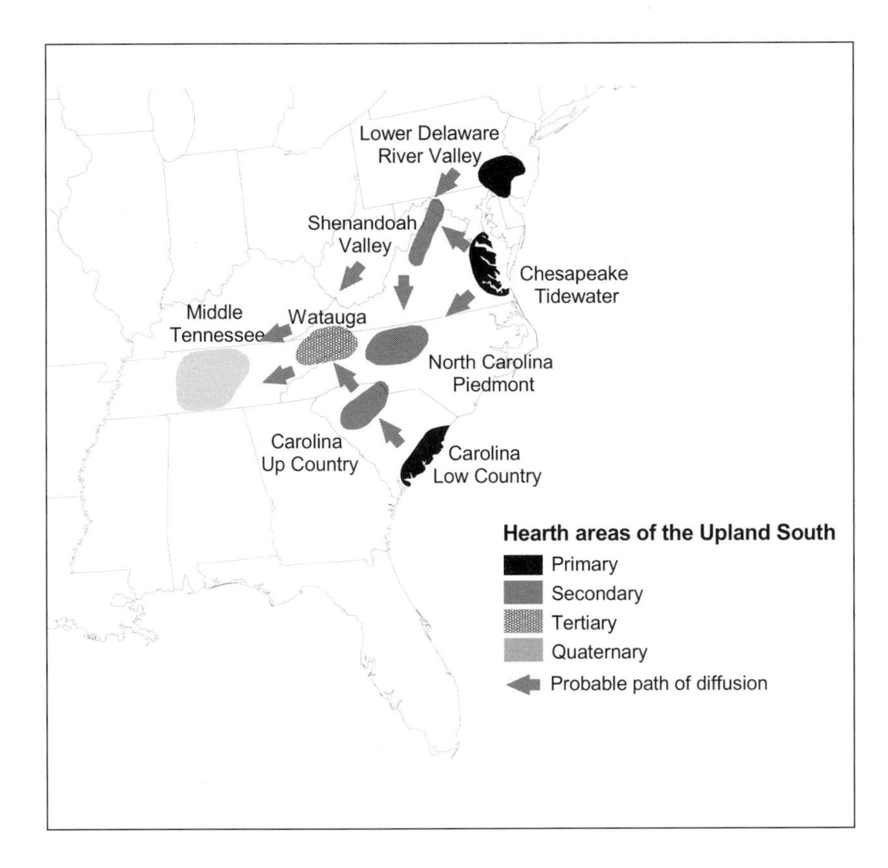

FIGURE 1.5 *Hearth areas of the Upland South. Middle Tennessee, the quaternary hearth, played the most crucial role, for it was there that the regional culture coalesced between about 1790 and 1815 and from there that it spread. (Sources: Mitchell 1978, 75; Ramsey 1964; Mitchell 1977; Lyle 1970–74 and 1972; Jordan 1993c, 111, 176, 190.)*

1770s. The final fermentation and coalescence of upland southern culture occurred in a quaternary hearth area, Middle Tennessee. The year was, say, 1810. Subsequently, natives of Tennessee would play the dominant role in the spatial expansion of upland southern culture. The Upland South, most immediately, is the child of Tennessee, and particularly of Middle Tennessee.

What Peoples?

Diverse hearths imply, correctly, a mixing of multiple peoples, a mixture finally completed in Tennessee. The main ingredients included English, Celts, Germans, American Indians, Swedes, and Finns, with a broad smattering of lesser groups.[22]

At the outset, we need to understand that a lot of nonsense has been written about the origins and cultural character of the upland southern population. They are decidedly not "Anglo-Saxons" or "Scotch-Irish" or even in the broader sense "Celts," as some have claimed.[23] Indeed, any notion of European ethnicity surviving into the quaternary Middle Tennessee hearth of the Upland South is nonsense. That, in fact, is exactly what Middle Tennessee and, to a certain extent, even Watauga was all about—the blending through intermarriage and acculturation of all the constituent groups into a single, distinct new population and culture. Nor did anything so bizarre as a "subconscious persistence" of Scotch-Irish identity occur in the Upland South.[24]

This does not mean that we need not analyze the ingredients of the proto-upland southern population. Each constituent parental group contributed elements to the emerging new highland culture, and the traditional upland southern way of life involved, in part, a sifting and winnowing process that preserved what was useful or appealing and discarded the rest.

Yes, the Scotch-Irish were abundantly represented, and virtually the entire Upland South stands out as a region of greater-than-average "Irish" ancestry in the United States censuses of 1980 through 2000 (Figure 1.6). Yes, undoubtedly the Scotch-Irish helped shape upland southern culture, though their Presbyterianism, burdened as it was by predestinarianism and the requirement for seminary-educated clergy, was crushed by the Baptists and Methodists. The Scotch-Irish legacy lives on in mountain music (with a screeching fiddle replacing the bagpipe), in whiskey-making, and in a hundred other ways. What does not survive is Scotch-Irish ethnicity or group identity. Relatively few upper southerners labeled their national origin as "Scotch-Irish" in the 1980, 1990, or 2000 national censuses—the only ones to include questions concerning remote ancestry. In the 2000 census of Texas, for example, only 338,000 persons, or 1.6 percent of those listing ancestry, claimed "Scotch-Irish." "American," "United States," or "unknown" proved to be far more common responses.[25]

In fact, "England" (or "English") was the most common first-listed origin reported by upland southerners in the censuses. All but two counties in Kentucky, for example, reported English ancestry pluralities, as did all Tennessee counties except a few with African pluralities in and around Memphis.[26] This "English" response reflects more the ignorance of upper southerners concerning their remote national origins and particularly their lack of knowledge of the basic geography of the British Isles. Again, we should reject the notion that upland southerners are Anglo-Saxons, though English ancestry and cultural legacy is undoubtedly common throughout the region. The square, "English-plan" one-room house is the most common folk dwelling throughout the Upland South (Figure 1.7).[27]

While not particularly numerous, descendants of the Finns and Swedes who settled the lower Delaware Valley hearth in the middle seventeenth century also contributed to the mixed upland southern population. They contributed diverse elements of the backwoods frontier forest colonization culture that survived well into post-pioneer times in the Upland South, including most notably techniques of notched-log carpentry and hunting methods.[28]

(opposite top)
FIGURE 1.6 *Counties with an above-average proportion of "Irish" national ancestry, 1980. The outline of the Upland South can be detected. (Source: Allen and Turner 1987, 50.)*

(opposite bottom)
FIGURE 1.7 *An English-plan, single-pen log house, in the Great Smoky Mountains of western North Carolina. (This postcard was published by W. M. Cline Company of Chattanooga, Tennessee, the copyright holder, and also bears the inscription "Tichnor Quality Views, Reg. U. S. Pat. Off., made only by Tichnor Bros., Inc., Boston, Mass." It is card no. 62017 and also has the number 277 printed on it. No successor to either Cline could be located, nor could the date of the photo be ascertained.)*

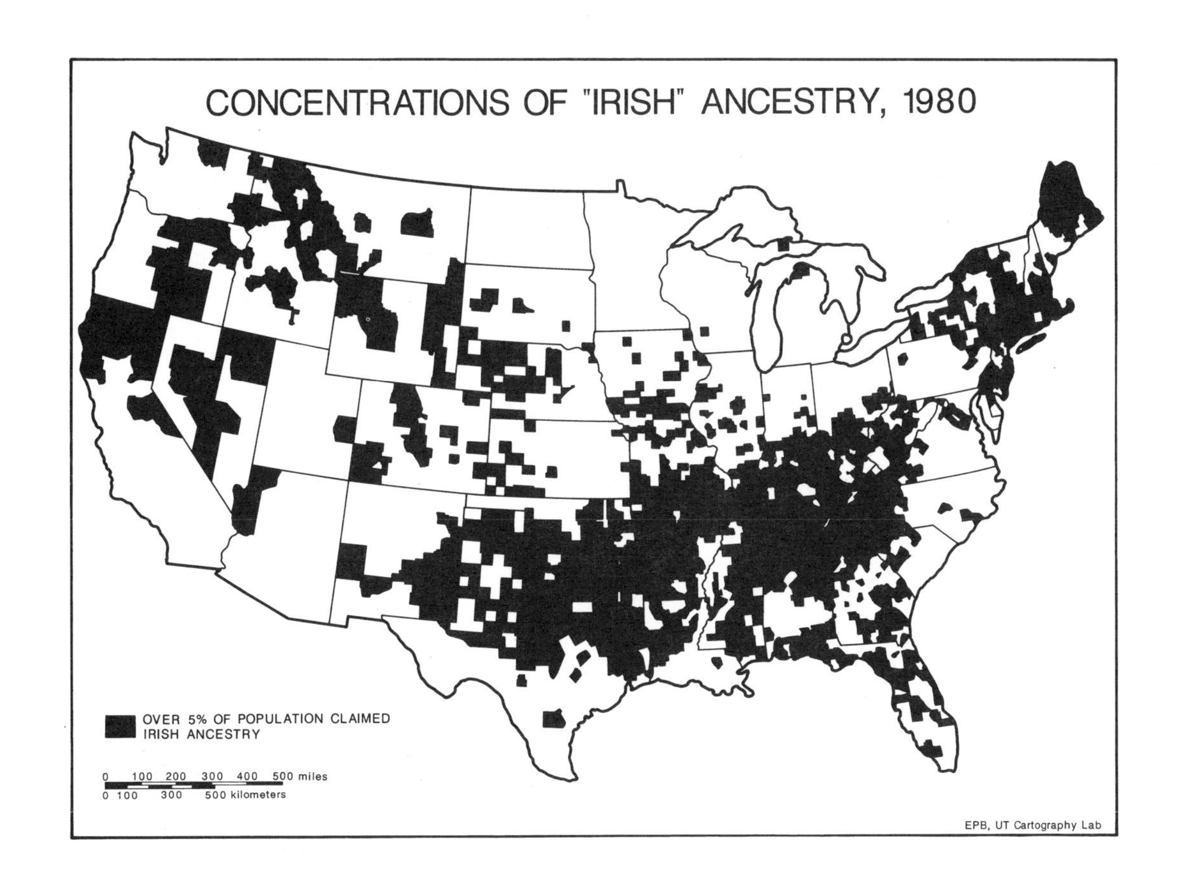
CONCENTRATIONS OF "IRISH" ANCESTRY, 1980
OVER 5% OF POPULATION CLAIMED
IRISH ANCESTRY
0 100 200 300 400 500 miles
0 100 300 500 kilometers
EPB, UT Cartography Lab

The Pennsylvania Germans

And what of the Pennsylvania Germans? The role of these "Dutch" was pervasive in much of Pennsylvania and south into the Shenandoah Valley secondary hearth, particularly in the post-pioneer period. Much of the Midwest bears their imprint. Their livestock-fattening agricultural system—with its huge, multi-level barn with a banked entrance to the upper level and projecting, cantilevered "forebay"—tells us where the "Dutch" went.[29]

Some of the Pennsylvania German descendants went on from the Shenandoah Valley to Watauga, and their surnames, corrupted in spelling, appear throughout the Upland South—Baugh, Coke, Wilbarger, Varner, Cline, Wise, Yokum, and many others. They added the yodel to country music, contributed the vernacular nickname "Bubba" to the hill southern vernacular, and in diverse other ways lent fragments of their culture to the Upland South.[30]

But the fact of the matter is that Pennsylvania German influence weakened southward. At the town of Buchanan, beyond Lexington and near Virginia's famous Natural Bridge, the southern reach of the Shenandoah Valley ends, pinched out by crowding ridges. Narrow defiles and lesser vales replace the fruited plain of Shenandoah there, and the Great Valley of the Appalachians only reappears in vigor more than 100 miles to the southwest, in Watauga and the Valley of East Tennessee. With a few exceptions, most notably around Winston-Salem, North Carolina, the Pennsylvania Dutch never migrated in any substantial numbers south beyond Shenandoah. To this day, the pinchout of the Valley beyond Lexington remains a German-British border of sorts. Watauga, Middle Tennessee, and the Upland South at large would bear relatively little German influence. In fact, the Upland South can be negatively defined as those parts of the Midland culture region where "Dutch" ancestry and influence remained minimal.[31] True, small German enclaves occur in some East Tennessee counties, but this is decidedly not "Dutch Country."

The Pennsylvania Germans declined to enter Watauga in force for several reasons. They were not favorably impressed by the poorer lands beyond the valley pinchout, and simultaneously the fertile lands of the lower Midwest beckoned them. Also, East Tennessee lay too remote from the urban markets of the Northeast to permit the "Dutch" livestock fattening system to take root and thrive. The Shenandoah Valley could and did become a stock feeder area, but distance dictated that Watauga remain a producer of lean "stockers," an enterprise scorned by the "Dutch." They would go west instead and establish the Corn Belt.[32] Their Teutonic dominance there, in the Midwest, would be massively reinforced by immigration directly from Germany in the nineteenth century, further distancing the Corn Belt culturally from the Upland South.[33]

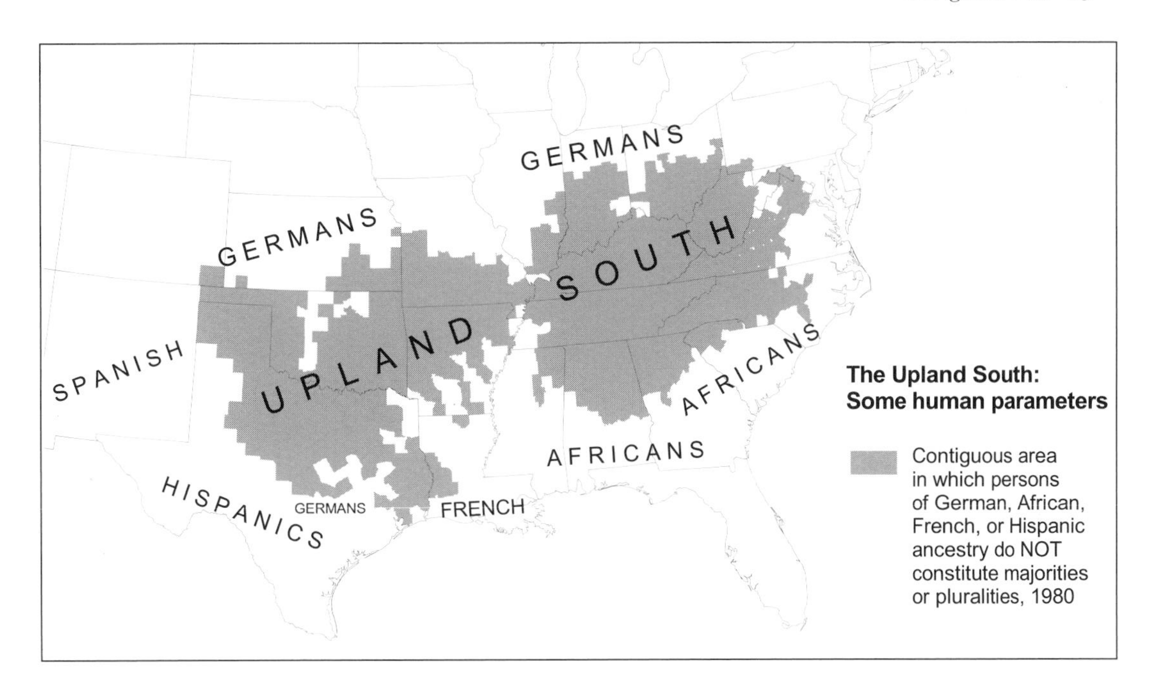

FIGURE 1.8 *Human parameters of the Upland South. (Source: Allen and Turner 1987, 54, 65, 145, 158, 166.)*

If the northern border of the Upland South is marked by German majorities and pluralities, the remainder of its perimeter also has an ethnic aspect. To the south and east, an African plurality or majority characterizes the coastal plain, while in Louisiana a French-derived population marks the end of the Upland South. To the west live Hispanic peoples, and in central Texas another German area provides the boundary (Figure 1.8).

Mixed Bloods

One additional and essential ingredient in the upland southern human mix deserves our consideration—the American Indian. The 1980 federal census revealed statistically a fact we should have known all along, namely that the Upland South reports the highest frequencies of *partial* Indian ancestry of any region of comparable size in the country (Figure 1.9). That is, a greater proportion of the population in the Upland South claimed Indian bloodlines, but without professing to be ethnic Indians or sacrificing their identity as Anglo-Americans. I have earlier described these people as the "Anglo Mestizos."[34] Upon questioning, they usually claim only a little bit of Indian blood, often confined to a temporally remote, almost invariably female, and often royal ancestor. "My great-great grandma was a Cherokee princess" would be a typical claim. I tend to believe these claims (shorn of royalty), and I feel sure that partial Amerindian ancestry is far more common in the Upland South than its people know.

Earlier censuses and research had revealed that persons belonging to the so-called "little races"—usually triracial mixes of Anglos, Blacks,

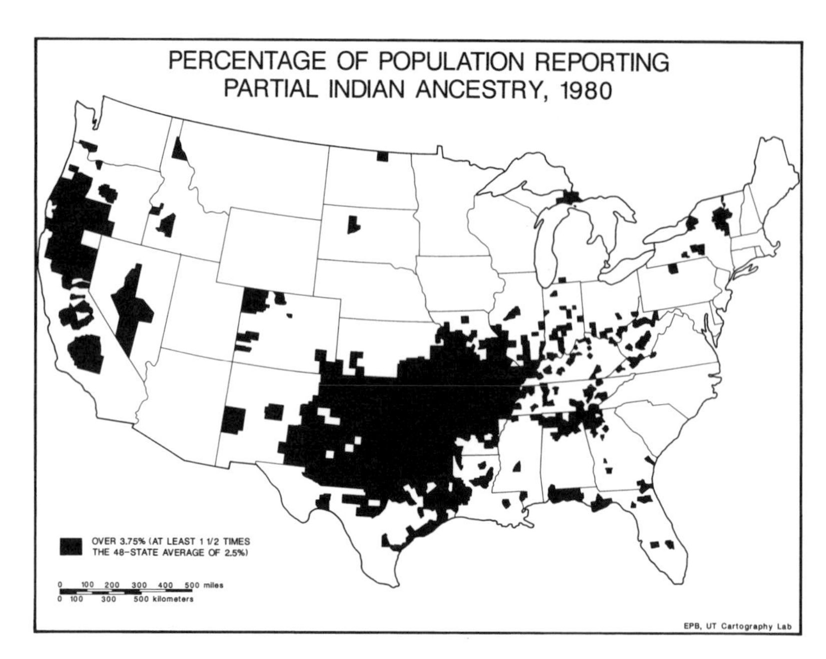

FIGURE 1.9 *Counties with an above-average proportion of partial American Indian ancestry. The great majority of such persons does not claim to be ethnic Indian, but instead to have some unspecified amount of Indian ancestry. Most ethnic Indians, in fact, are excluded from this response. Here, again, the geographical dimensions of the Upland South can be detected. (Source: simplified from Jordan 1993b, 177.)*

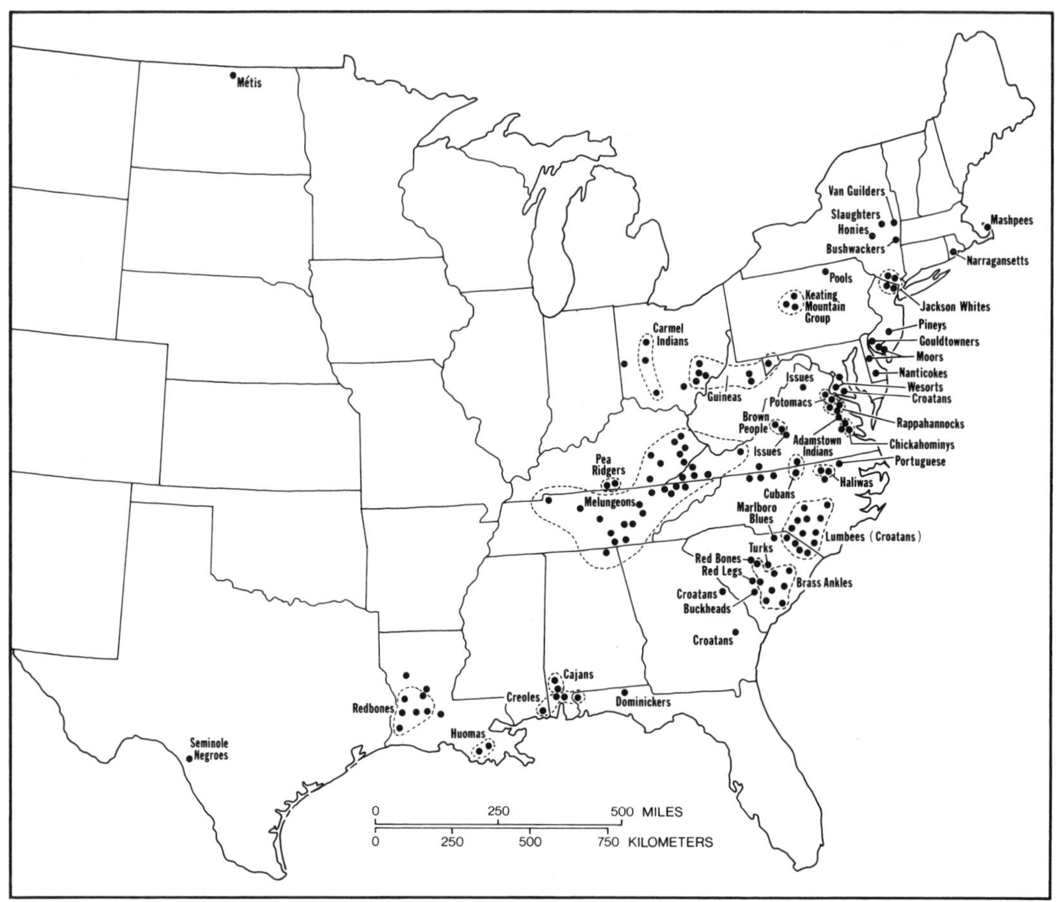

FIGURE 1.10 *Ethnic/racial isolates of reputed partial American Indian origin. One of the most widespread such groups in the Upland South are the Melungeons. If they and the "Pea Ridgers" are added to the pattern shown in Figure 1.9, the eastern half of the Upland South joins the western part as a single region having considerable Indian blood. (Source: Jordan 1993b, 181.)*

and Indians, or simply mestizos with a bit too much Indian blood—also are more common in the Upland South (Figure 1.10). The most widespread such group, the Melungeons, occupy parts of three different upland southern states.[35] Moreover, many remnants of ethnic tribal Indians survive in the Upland South, most notably but not exclusively in eastern Oklahoma, and these are usually mixed bloods, too.[36]

Mixing with the forest tribes of the eastern United States introduced many vivid elements of Amerindian culture into the upland southern way of life. The Native American influence is revealed in foodways—in particular, a fondness for cornmeal products; in dialect, especially loanwords; in folk religion such as the snake-handling of certain Appalachian sects; in dance; in folktales; in forest lore; in ethnobotanical medicine; and more.[37]

Tennessee Extended

These multiple peoples in time blended into one and spread vigorously, expanding the domain of the Upland South.[38] In this expansion, Tennessee played the dominant role, just as it had earlier served as the most immediate hearth of the upland southern way of life. Decade after decade, the daughters and sons of Tennessee spilled out of their native land, creating a diaspora that largely accounts for the broad geographical extent of the greater Upland South. Tennessee not only culturally mothered, but also peopled the Upland South (Figure 1.11).

As early as 1850, nearly a quarter-million native Tennesseans lived outside the boundaries of the state.[39] So significant was their exodus that the editor of the *Historical Atlas of the United States* saw fit to

FIGURE 1.11: *The Tennessee diaspora. Any county is included in which Tennesseans were the largest or second largest out-of-state group, by birth and/or prior state of residence, in the censuses of 1850 through 1880. (Sources: manuscript U. S. census population schedules; Jordan 1969, 87; Gerlach 1976, 178–180; Gerlach 1986, 23, 51–76; Meyer 2000, 289; Meyer 1976b, 5; Meyer 1976a, 158; Rose 1985, 206–207; Kerr 1953; Lathrop 1949; White 1948; Treat 1967, 252-259; Walz 1952, 34-35; Walz 1958, 77–116; Garrett 1988, 49.)*

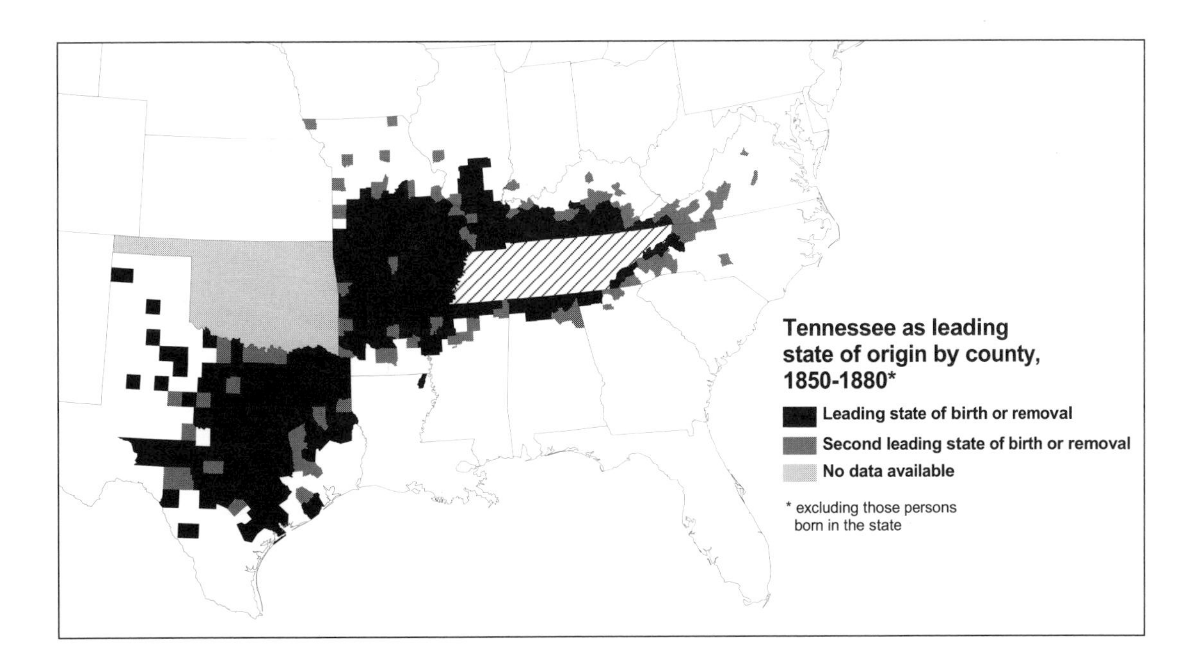

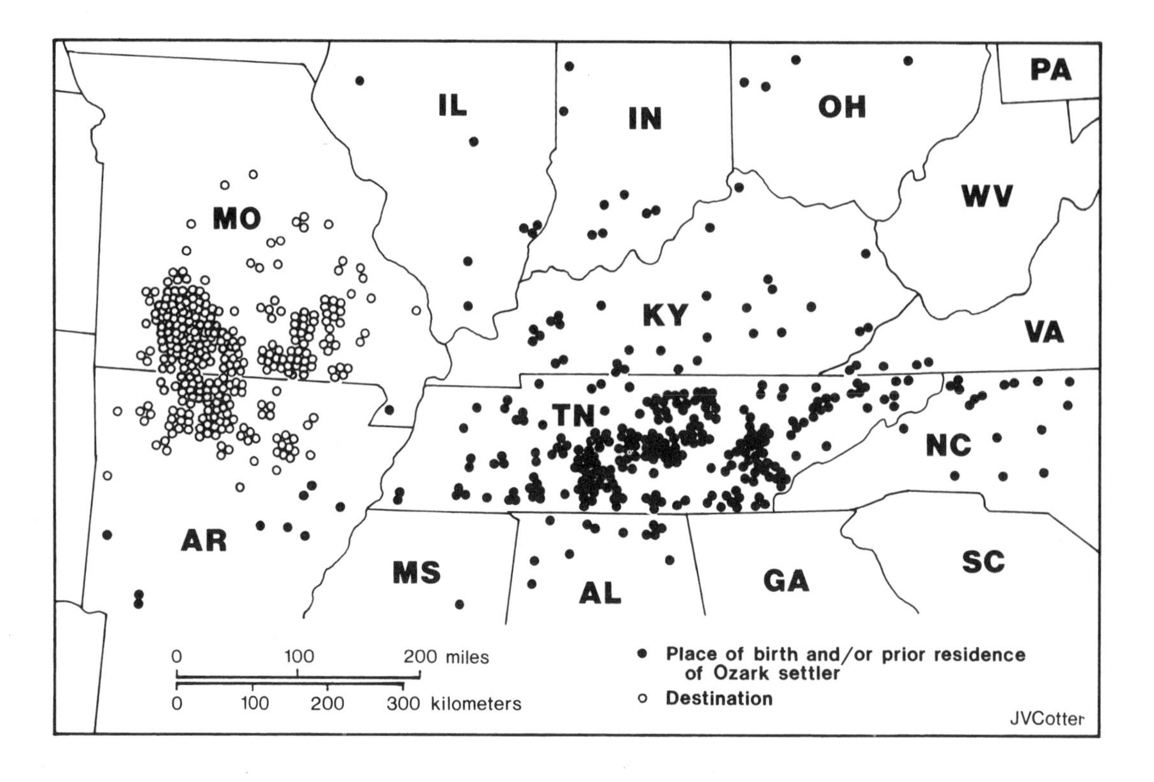

FIGURE 1.12 *Sources and destinations of selected settlers in the Ozark Mountains, 1820–1860. The source dots represent both places of birth and prior residence, and the same person may, as a result, be represented more than once among the sources. The importance of Tennessee, and especially Middle Tennessee, is evident. Compiled from selected pioneer reminiscences. (Source: adapted and simplified from Clendenen 1973.)*

include a map revealing the huge area west of the Mississippi River where Tennesseans formed the largest nativity group.[40]

Proximity, combined with the dominantly westward latitudinal flow of migration, dictated that Arkansas and Missouri would receive the largest contingents of Tennesseans. Close to one-third of all expatriate Tennesseans resided in those two states by 1850.[41] Arkansas has been called a "child of Tennessee," and the *Missouri Historical Review* long ago published an article entitled "Missouri's Tennessee Heritage."[42] Middle Tennessee, in particular, contributed heavily to the Ozarks, and some 31,000 square miles of that mountainous area became a contiguous area of Tennessean colonization (Figure 1.12).[43]

Nor did the Old Northwest fail to share in this migration. Illinois ranked third, behind Missouri and Arkansas, accounting for 13.4 percent of all former Tennesseans in 1850.[44] Indeed, southern Illinois was early and massively influenced by immigrants from the Volunteer State, forming the basis of that state's "Sucker" population and prompting an article entitled "The Influence of Tennesseans in the Formation of Illinois."[45] Doug Meyer, in a long series of scholarly articles, has illustrated the Tennessee presence in Illinois. More than one-quarter of all immigrants to the Shawnee Hills of far southern Illinois came from Tennessee.[46] Indiana received a far weaker impulse from Tennessee, and Ohio even less. These two states admittedly weaken the notion that the Upland South is simply Tennessee Extended. Only 13,000 natives of

Tennessee resided as "Hoosiers" in Indiana in 1850, accounting for 5.3 percent of the former Tennesseans, and in no county did they form the largest state nativity group (Figure 1.11).[47]

Kentucky, owing to its proximity, received a huge input of Tennesseans, especially in the southern half of the state (Figure 1.11). While their input was preceded and mitigated by immigrants from Pennsylvania and Virginia, Tennesseans can be said to have been primarily responsible for the upland southern character of much of Kentucky.[48]

The coastal states of the South differed greatly in the degree of Tennessean influence and presence. North Carolina, Georgia, Alabama, and Mississippi saw Tennessee immigration limited largely to their western or northern tier of counties, and Louisiana was, oddly, scarcely touched (Figure 1.11). Even so, Tennessee ranked as the second leading birth state of immigrants to Mississippi and in 1850 accounted for 18.3 percent of the free population born out-of-state.[49] Anglo-Texas, by contrast, became another Tennessee Extended, particularly in its formative years before 1850. A huge swath of Texas was populated by natives and/or former residents of Tennessee, including influential persons such as David Crockett and Sam Houston. Tennesseans established a major presence in Texas as early as the 1820s, and among them were venerable surnames such as Holston. A far disproportionate number of Tennesseans fought for Texas independence at the Alamo and San Jacinto (Figure 1.13).[50]

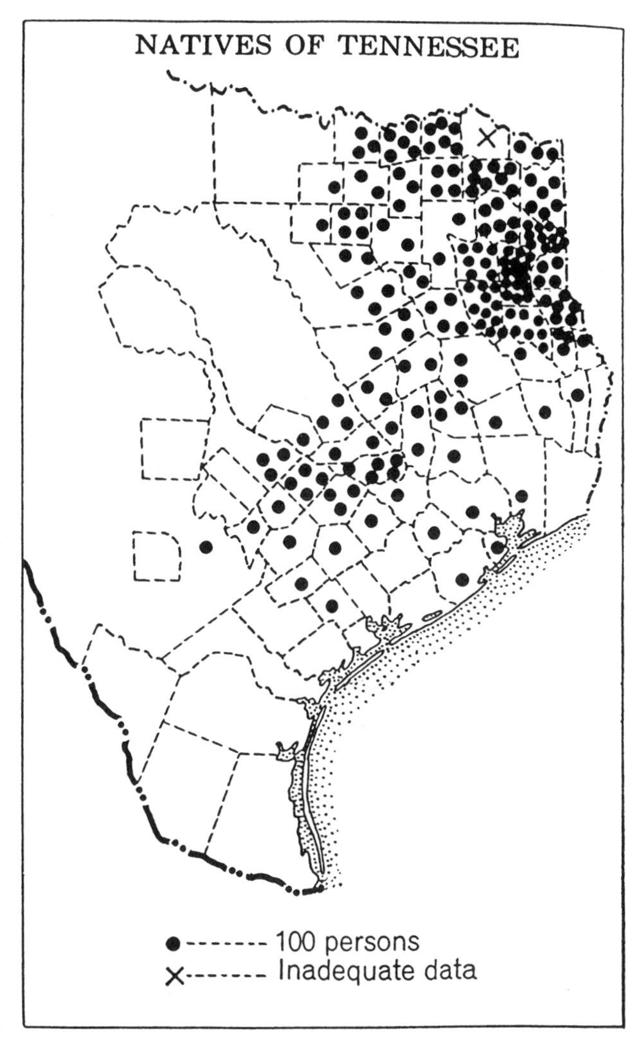

FIGURE 1.13 *Distribution of natives of Tennessee in the settled, eastern half of Texas, 1850. Notice how widespread they were. (Source: adapted from Jordan 1969, 87.)*

Nor did the reach of Tennessee Extended end there. Even into southern and eastern Kansas, individual townships had pluralities of native Tennesseans. One Chase County Township reported 27 percent of its immigrant population as Tennessee-born in the state census of 1865. An upland southern cultural presence, if muted, accompanied the Tennessee settlers into this part of Kansas.[51]

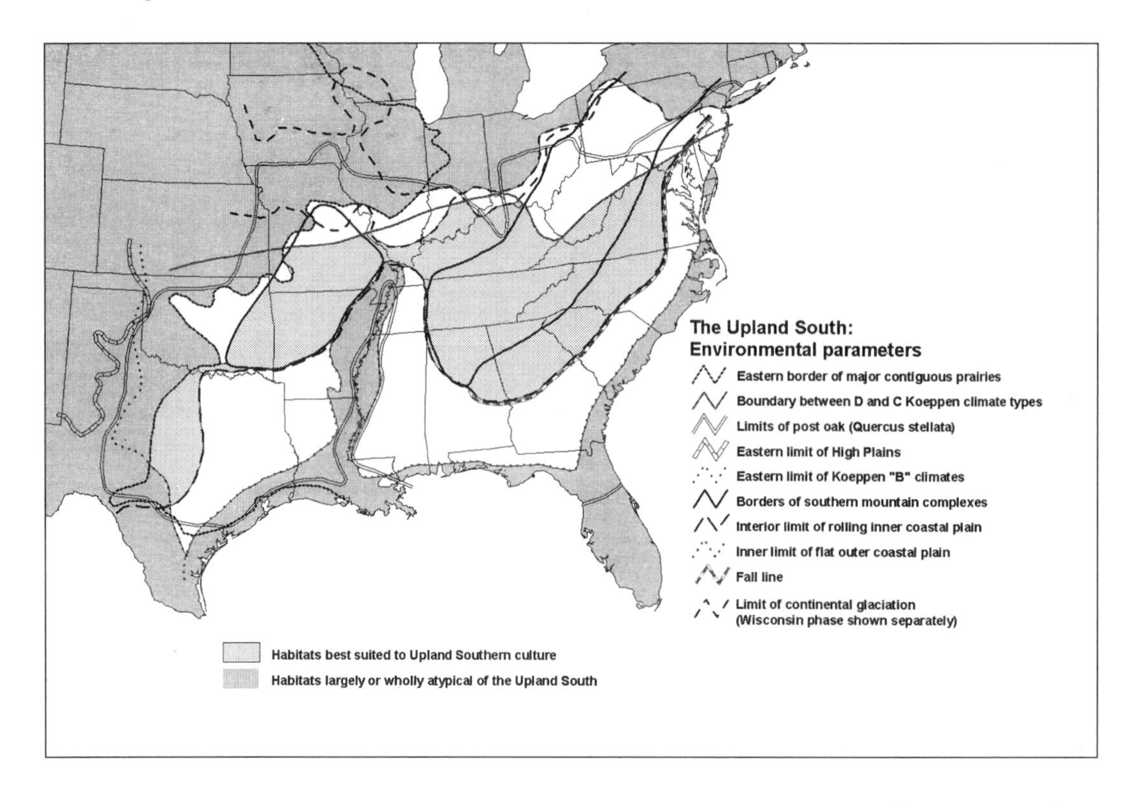

FIGURE 1.14 *The Upland South: some environmental parameters. (Sources: Hudson 2000, 14, 67, 68, 70–71, 74–75; Transeau 1935; Jordan 1973a; Dicken 1935; Wilhelm 1972, 113; Jordan 1964, 209–211.)*

Highland Habitat

The expansion of the Upland South from its Tennessee hearth largely involved occupying a specific type of habitat. Environmental as well as cultural parameters came into play in upland southern ethnogenesis and diffusion. Broken terrain provided one key element of the upland southern habitat, but only one among several (Figure 1.14). The mountainous, hilly, and rolling lands of the southern Appalachians and Piedmont, bordered on the east and south by the Fall Line and Atlantic/Gulf coastal plain, provided the initial niche for expansion. This habitat bridged across the narrows in the Shawnee Hills of southern Illinois to the Ozarks and Ouachitas, a western twin, offering the same dissected plateau, parallel ridges, and elongated central valley as the Appalachians. Passing across another hilly narrows, the upland southern folk reached the Hill Country of central Texas, where another familiar habitat awaited them.[52] Nor did the expansion end there. Decades later another montane region, far to the west, beckoned. Appalachian hill folk poured into the Coastal Range and Cascades of western Washington, though that distant settlement zone cannot seriously be considered part of the Upland South.[53]

Most of the upland southern expansion occurred in the mixed hardwood forest of the South. Perhaps the single best indicator of this preferred floral habitat is the post oak tree (*Quercus stellata*). The Linnean adjective here means "star-shaped," a reference to the leaf. In their niche-filling diaspora, the post oak provided settlers their star of Bethlehem (Figure 1.14). They at once utilized and venerated it, often leaving mature post oaks standing around their houses and in their graveyards, possibly because some saw its star as a cross.[54] The hardwood forest also included diverse other types of oak, as well as hickories, black walnuts, and poplars.

We need also to understand that the upland southern habitat is "humid subtropical," lying in the Koeppen climate type labeled "Cfa." The Russian geographer Leo Gumilev long ago proposed that ethnogenesis—the emergence of new and distinctive cultural groups—most often occurred along ecotones, or habitat boundaries. The tertiary and quaternary hearth areas I propose for the Upland South all lie near or on the ecotone between subtropical and continental climates—the "C"/"D" border in the Koeppen system (Figure 1.14).[55] We should not underestimate this environmental influence in the formation of the upland southern way of life. In this sense, the Upland South is partly the result of Pennsylvanians entering the subtropics.

The habitat sought by migrating upland southerners can also be defined negatively. They generally rejected glaciated lands; the cold or dry climate types, labeled types B and D in the Koeppen classification system; wetlands; pine forests; open grasslands, or "barrens" as the hill folk called them; and flat terrain, such as the southern outer coastal plan, the Delta country along the lower Mississippi River, and the High Plains of western Texas, Oklahoma, and Kansas (Figure 1.14).[56] In the process, they often chose lands inferior in agricultural potential to the ones they shunned. In sum, they selected the humid Koeppen type Cfa climate; forested lands, usually oak-dominated; and well-drained, unglaciated, hilly to rolling terrain. If forced by population pressures to move into regions of a significantly different physical character, they generally failed to implant their distinctive culture.

Ethnogenesis

My underlying hypothesis is that a distinctive upland southern culture and way of life arose not just in particular hearth areas among a particular mix of peoples and spread through a particular habitat, but that multiple processes worked to bring about this ethnogenesis. Preexisting elements of Euroamerican culture were selectively and archaically retained, modified or magnified in importance, or rejected. Invention occurred, if rarely. In addition, indigenousness occurred—borrowings from the Indians of the Upland South, as earlier suggested. The result of these processes was a regional culture—one of several in the North American heartland—easily distinguished from its Lowland coastal southern and Midland Pennsylvanian parents, and as well from its Midwestern sibling.

Much was retained from the Pennsylvanian culture, perhaps most importantly notched log construction, certain folk house and barn plans, and the county form of local government administered from courthouses in county seat towns. Often, even typically, upland southerners retained elements of Midland folk culture such as notched log construction long after they disappeared in their source region, with the result that much of the distinctiveness of the Upland South lies in its archaic Pennsylvanian character. Likewise, borrowings from the Tidewater and Lowland South often survived longer among upland southerners. Tobacco remains a hill crop today more than a century and a half after it disappeared from the Virginia Tidewater, and free-range Carolina "cowpen" herding persisted into the modern era in the Upland South, long after it succumbed to pine plantations and peanut farming on the southern coastal plain.

At the same time, much from the parent cultures failed to find a place in the upland southern way of life. The German-inspired "continental" type of folk house, featuring three rooms grouped around a central oven hearth, never reached the Upland South, nor did the equally German "Pennsylvania" barn, a massive multi-level structure with a banked ramp entrance to the upper level, which also features a projecting, cantilevered "forebay." Black slavery and the plantation system similarly found scant welcome in the southern mountains, prompting Kentucky, Missouri, and West Virginia to join or remain with the union.[57]

Certain traits present in the parent cultures achieved a magnified importance in the Upland South, beyond the level they had achieved back in Pennsylvania or the coastal South. This magnification added to the uniqueness of the Upland South. Two folk houses provide examples. The "dogtrot" house, consisting of two main rooms with an open breezeway between, was introduced into Pennsylvania's Delaware Valley in the 1600s, but never became a widespread type in the Pennsylvania culture area (Figure 1.15). A comparably ancient introduction brought the "saddlebag" house to the coastal plantations of Virginia and South Carolina. Outwardly similar to the dogtrot, the saddlebag house also has two main rooms, separated by a chimney instead of a breezeway (Figure 1.16). A servants' quarters, the saddlebag never became the dominant type of slave cabin. By contrast, both the dogtrot and saddlebag house types achieved major importance in the Upland South.[58] (Read more about them in Chapter 3.)

Some other parental traits underwent substantial modification in the Upland South, sufficient to create distinctive regional types. For example, the Pennsylvania town square, or "diamond," featured avenues

FIGURE 1.15 *Dogtrot house, Newton County, East Texas. (Photo by the author, 1978.)*

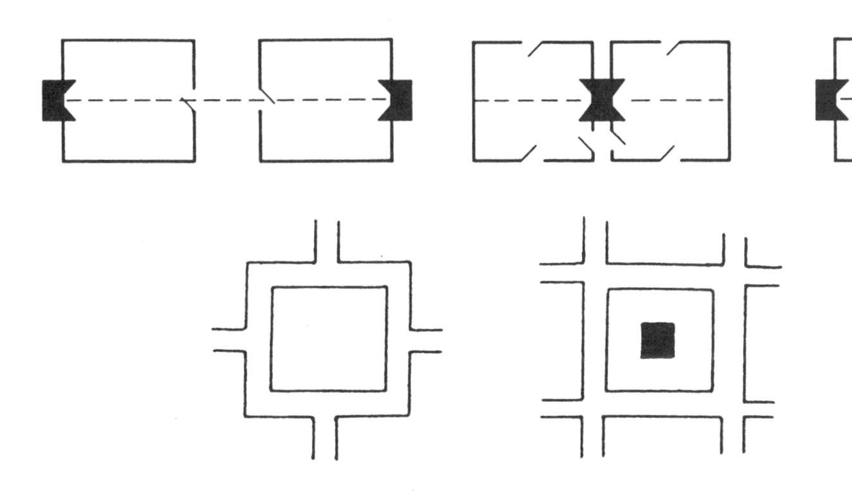

(middle)
FIGURE 1.16 *Three types of upland southern double-pen log houses: left = "Cumberland;" center = "saddlebag;" and right = "dogtrot."*

(bottom)
FIGURE 1.17 *Two types of courthouse squares: left = "Shelbyville," with the courthouse in the center, and right = the Pennsylvania town diamond, in which the courthouse is not in the center. The diamond is also called a "Philadelphia" square. If, as sometimes happens in the Upland South, a courthouse stands in the middle of the diamond, it is called a "Lancaster" square.*

entering at the four midpoints of the square. In Tennessee, a modified type, the "Shelbyville" plan, in which avenues entered at the corners of the square and the courthouse stood in the middle, arose to become dominant throughout most of Tennessee Extended (Figure 1.17).[59] The most important upland southern folk barn, the "transverse-crib," likewise involved a modification of an older Pennsylvanian type.[60] (The transverse-crib barn and Shelbyville square will provide the focuses for Chapters 4 and 5.)

In some cases, modification was of such a magnitude as to approach the status of true innovation. But, in fact, invention remained rare in the formation of the upland southern regional culture. Culling played a larger role than invention, perhaps mainly because the several parent cultures, when combined, possessed more diversity than was needed or wanted.

Methodology

I do not pretend here to write a definitive cultural geography of the Upland South. Both my ability and ambition demand less. Instead, I propose to sample a few items of folk culture and to employ only one methodology. The list of what I will *not* consider is far longer than what I include. You will find no consideration of such diverse subjects as dialect, folk medicine, food, drink, oral history, religious faith, music, crafts, or the modern era, among many others.

The methodology I employ throughout the book is an analysis of the traditional *cultural landscape*—the built environment or greater artifacts. From this folk landscape, I feel, we can learn a great deal about the Upland South, about its character, diffusion, and geographical extent. "Reading" the cultural landscape has long remained a leading methodology for cultural geographers, and with good reason. People reveal a great deal about themselves in their greater artifacts—houses, barns, town plans, graveyards, church structures, fences, and the like. People often lie, if unintentionally, but their artifacts, if correctly read and interpreted, invariably tell the truth. We will begin with the work of folk carpenters.

CHAPTER 2

The Hill Southern Log Culture-Complex

FOLK CARPENTRY, reflecting the handiwork of common people, contains diverse diagnostic clues concerning a culture, its origin, and its spatial diffusion. In the Upland South, this carpentry involved notched-log construction, a technique once so dominant throughout the region as to prompt the term "log culture-complex."[1]

Pervasiveness and Persistence

Notched logs provided the walls of houses, barns, cribs, smokehouses, fences, chicken coops, chapels, schools, jails, and just about every other kind of structure raised by upland southerners (Figure 2.1).[2] Derived from the Pennsylvanian parent, upland southern log carpentry exhibited colonial continuities, indigenous selections, and modifications that collectively lent it a special regional character. The carpenters utilized a small number of rather simple techniques, easily described and listed. Logs could be left in the round, as was typically the case on early pioneer cabins and many later outbuildings, or else hewn on two sides with adz and ax to produce flat-faced walls.

Whether round or hewn, the logs touched only at the notched corners, leaving open cracks, or "chinks," between the timbers that required stopping up with clay or some other filler to allow a tight wall. Most often the hewing produced a beam of moderate thickness, but upland southern carpenters sometimes "planked" logs to a narrow width (Figures 2.1 and 2.2). In either method, the log retained the score marks of the ax even after adz hewing.[3]

Several characteristics lent distinctiveness to upland southern log construction, differentiating it from the Pennsylvania type. Perhaps, most notably, this carpentry survived far longer in the Upland South

(top left)
FIGURE 2.1 *Half-dovetail notching on a log house in Sequoyah County, in the Cherokee area of Oklahoma. The wood is post oak, the favorite upland southern tree for log building. (Photo by the author, 1981.)*

(top right)
FIGURE 2.2 *Half-dovetail notching near Cherokee, Swain County, North Carolina. The logs have been "planked" to a thin profile, allowing two different planks to be taken from each log. (Photo by the author, 1980.)*

than elsewhere in the eastern half of the United States, a persistence that lent an archaic quality to the regional cultural landscape.[4] Decline of the tradition eventually set in, but the cultural landscape of the Upland South retains tens of thousands of log buildings even today.[5]

Corner Notching

The structural key to such carpentry lay in the corner notch, because it both bore the weight of the entire structure and locked the logs laterally, so they could not slip sideways out of place. While logs can be notched in diverse ways, only a few notch types were employed by upland southern carpenters (Figure 2.3). The technology remained simple, lacked complexity, could be relatively easily learned, and required few tools—mainly the ax, adz, and saw.[6] It possessed the same archaic quality of the log culture-complex at large. Eliot Wigginton and his students found carpenters in northern Georgia still able to fashion the few traditional upland southern notches in the 1960s and 1970s.[7]

Among the notch types were the half-dovetail, "V", saddle, square, and diamond types (Figures 2.1, 2.3, and 2.4). Only these five found a place in the upland southern carpentry repertoire, and among them only the first four listed enjoyed widespread usage. Even the V-notch, the most common Pennsylvanian type and widely used by the Germans of that state, gained only a rather minimal acceptance in the

(top left)
FIGURE 2.3 *"V"-notching on a post oak log house in Cooke County, North Texas. The V-notch is less common in the Upland South than half-dovetailing and is more closely associated with the Pennsylvania Germans (even though it is of Scandinavian origin). V-notching appears consistently as a minority type in the Upland South. (Photo by the author, 1972.)*

(top right)
FIGURE 2.4 *"Square" notching, Conway County, Arkansas. This type of notch, though widely encountered, is less common than V-notching in the Upland South and is more closely associated with the coastal plain of the South. (Photo by the author, 1988.)*

(bottom right)
Figure 2.5 *"Saddle" notching, Hood County, Central Texas. This notch occurs mainly on outbuildings in the Upland South and is not diagnostic of the region. (Photo by the author, 1975.)*

Upland South, reflecting the weakness of "Pennsylvania Dutch" settlement in the region. It would become more important in the Midwest than the Upland South. Saddle notching was reserved for outbuildings consisting of unhewn logs (Figure 2.5), though the V-notch could also be used on round logs. The square notch appears far more commonly in the coastal plain of the South than in the uplands, and the diamond notch occurs with frequency only along the axis of the Fall Line in North Carolina.[8]

Half-Dovetailing

One log notch type—the half-dovetail—reigned as the dominant one throughout most of the Upland South (Figures 2.1 and 2.2).[9] The region is so closely identified with common usage of this particular notch that I will use half-dovetailing as diagnostic of the Upland South, as a cultural landscape index of the degree of hill southern influence in any locality.[10] It provides the first of five diagnostic landscape elements I have chosen for emphasis in this book and is the one selected to represent the log carpentry tradition.

To fashion a half-dovetail notch, also called a *mitre* dovetail, the carpenter first used an ax to cut a 25° to 45° slope on the top side of an end of a hewn or planked log, sloping downward from the side of the log that was to face the inside of the room. The sloped surface was smoothed with short, careful ax strokes. Then another shaped log was placed atop the first, at right angles to it and forming part of the adjacent wall. Using a ruler about one inch in width, resting on the previously cut slope, the carpenter drew a line or made a scratch along the top side of the ruler on the butt end of the second log. Next he placed the ruler vertically, resting flat against the hewn interior wall surface of the first log and drew a vertical line to intersect the sloping line on the second log, a process then repeated on the opposite side of the second log. By cutting to these lines with the ax, the "corner man" produced a perfect locked fit between the logs. The lines could also be drawn by using a pattern cut to the desired slope angle. In either case, the width of the chinks between logs could be determined by the amount of wood removed in the notch-cutting process.[11]

In most cases, the notching was sawn off flush to produce a "boxed" corner (Figure 2.1). Some carpenters instead allowed the half-dovetailed logs to project an inch or so beyond the corner, creating an "overhanging" notch (see Figure 3.3 in next chapter).[12]

All of this is difficult to capture in words, though many authors have tried. Put differently, in fashioning the half-dovetail notch the top side of the log is splayed or "the head of the notch slopes upward," while the "bottom edge remains flat."[13] That is, "the bottom of the notch is flat rather than sloping."[14] The reader is best advised to look at the end product in trying to visualize the process (Figure 2.1), or else refer to the splendid photographic sequence of the notch being cut presented in Wigginton's *Foxfire Book*.[15]

FIGURE 2.6 *The full-dovetail notch. While it is the parent of half-dovetailing, the full-dovetail is virtually absent from the Upland South.*

Modification

The half-dovetail notch is a visible index to the Upland South not only because it achieved greatest usage there, but also because it most likely involved a modification, or, more exactly, a simplification of a Pennsylvanian notch. The parent type, the *full-dovetail* notch, was more complicated, having sloped cuts both on the top and bottom of the joint (Figure 2.6). The modification not only made the notch somewhat easier to fashion, but also enhanced its ability to drain rainwater to the outside of the joint, prolonging the life of the log structure. (Compare figures 2.1 and 2.6.)

Originally, the parent full-dovetail notch came with the Swedes to the Delaware Valley in the 1600s (Figure 2.7).[16] Reintroduced by eighteenth-century German immigrants, it became a common Pennsylvania "Dutch" type.[17] The full-dovetail rarely accompanied Midlanders as they departed Pennsylvania, with the notable exception of the Shenandoah Valley. One rarely finds it in the Upland South.

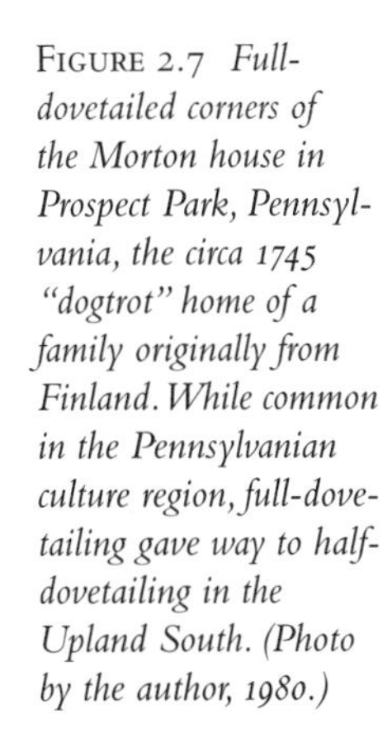

FIGURE 2.7 *Full-dovetailed corners of the Morton house in Prospect Park, Pennsylvania, the circa 1745 "dogtrot" home of a family originally from Finland. While common in the Pennsylvanian culture region, full-dovetailing gave way to half-dovetailing in the Upland South. (Photo by the author, 1980.)*

The half-dovetail type of joint occurs on log buildings in northern and central Europe, though infrequently.[18] It also occurs in European cabinet-making. As Henry Glassie said, possibly "it was a part of the full-dovetail tradition brought from Europe," but more likely the half-dovetailing of the Upland South represents an Americanism, a reinvention not derived directly from Europe or even from Pennsylvania itself. The half-dovetail modification possibly occurred in the Shenandoah Valley or adjacent Alleghenies, one of the secondary hearths of the Upland South (Figure 2.8). It occurs as a minority type in the Valley, as well as in the adjacent Blue Ridge on the east and in the ridge-and-valley region just west of the Shenandoah, on both sides of the West Virginia/Virginia border. Still, two early examples of half-dovetailing occur in Bucks County, Pennsylvania, just north of Philadelphia.[19]

Estyn Evans, without any supporting evidence, attributed the half-dovetail simplification to the Scotch-Irish.[20] He was probably right, if for no other reason because I know of not one single case where Pennsylvania Germans employed half-dovetailing, in the Shenandoah Valley or anywhere else. Instead, the "Dutch" almost exclusively used their ancestral full-dovetailing or else V-notching, which they borrowed from the Finns and Swedes in colonial times in the lower Delaware Valley.[21] But the carpenter who developed the half-dovetail need not have been Scotch-Irish. Bucks County had relatively few Ulster settlers and the Shenandoah Valley early housed a mix of peoples, including acculturated Swedes, Finns, Welsh, Tidewater English, and even some "Holland Dutch" from the Hudson River valley in New York.

Distribution

While likely developed in or on the hill fringes of the Shenandoah Valley, the half-dovetail notch did not become dominant there. One must move southwestward down the Great Valley of the Appalachians into the Watauga country in East Tennessee to find the first area where half-dovetailing is consistently the most common notch on log houses and barns (Figure 2.8). But just to the south, in the Valley of East Tennessee and in the nearby coves in the Great Smoky Mountains, half-dovetailing accounts for nearly two-thirds of all log houses.[22]

Western North Carolina also belongs to this earliest region of half-dovetail dominance (Figure 2.8). Perhaps the main route of southward diffusion ran through the Piedmont rather than the Great Valley. But the route between place of invention and region of acceptance does not matter. We need only know that the half-dovetail notch and the Upland South came together in East Tennessee and diffused from there into Middle Tennessee, where the degree of dominance of half-dovetailing rose to between 75 and 80 percent.[23] The half-dovetail notch

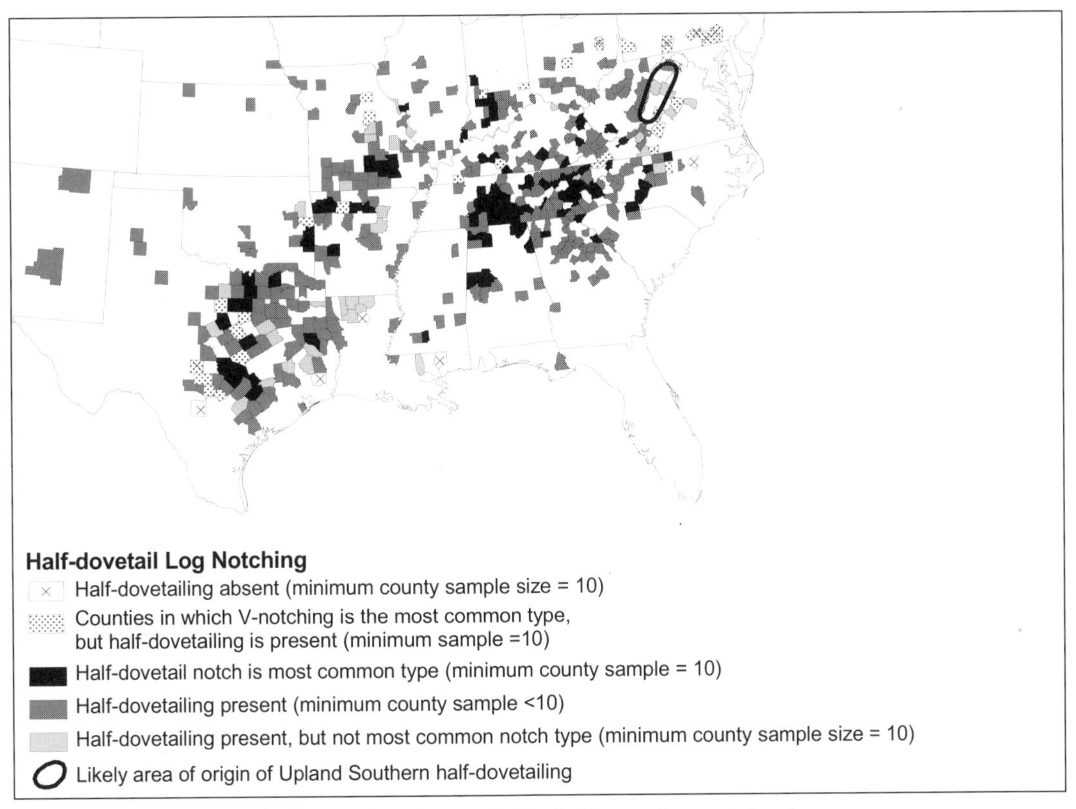

FIGURE 2.8 *Distribution of half-dovetail notching in the South. Because the map is based upon field observations by the author and numerous other persons, it should not be regarded as complete. Some counties were not inspected at all. (Sources: Martin 1989, 148; Stuck 1971, 228, 232; Glassie 1965a, 51; Milbauer 1996–97; Gavin 1995; Roberts 1985, 66; Lyle 1970–74 and 1972; Montell and Morse 1976; Elbert and Sculle 1982; Hutslar 1977 and 1986; Marshall 1981; Clendenen 1973; Morgan 1990; Martin 1984; Jordan 1978, 55; Kniffen and Glassie 1966, 61; Wilson 1975; Rehder 2002; and other sources cited in Jordan, Kaups and Lieffort 1986, 22-23; Jordan and Kilpinen 1990, 8.)*

illustrates the role of the Middle Tennessee quaternary hearth of the Upland South, where the culture coalesced into its final form.

From that base it spread with the Tennessee diaspora through virtually all of the extended Upland South. What is perhaps most remarkable is that, even on the outer fringes of the Upland South, half-dovetailing retained its dominance. In southern Indiana, it accounted for 74 percent of the 470 log buildings, both house and barns, inspected by Warren Roberts, and in eastern Kentucky for 67 percent of those included in Charles Martin's sample.[24] Near the confluence of the Illinois and Mississippi rivers, in Jersey County, Illinois, half-dovetailing still prevails on the very borders of Yankeedom.[25] And in distant North Texas, in the district called the East Cross Timbers, this type, locally called the "Missouri notch," accounts for three-quarters of all log dwellings.[26] Statewide in Texas, half-dovetailing is the most common type of log houses, perhaps not surprising when one considers that Tennessee was the leading state of birth of Anglo-Texans. The only exception would appear to be the Cherokee country of northeastern Oklahoma, where half-dovetailing is relatively rare.[27]

By contrast, in states which received few Tennesseans, such as Ohio, half-dovetailing appears less commonly.[28] To generalize greatly, we could say that dominance of half-dovetailing identifies the Upland South and is bordered on the north by a Pennsylvania Dutch-derived belt of V-notching across the lower Midwest and on the south and east by a zone of square-, saddle-, and diamond-notching in the coastal plain of the South (Figure 2.8).[29]

Things are, of course, never that simple. Half-dovetailing also existed in a completely independent log-building complex in North America. Rooted in early colonial northern New England and derived from a British military garrison house construction tradition, this separate notched-log complex spread into Quebec, Ontario, and the upper Midwest. Some half-dovetailing I have seen in northern Pennsylvania—the Yankee-settled tier of counties—belongs in this tradition rather than that of the Midland/Upland South. Fortunately, a belt devoid of half-dovetailing clearly separates the two areas of occurrence, so that the definitive character of half-dovetailing in upland southern carpentry survives.

The most essential expression of log carpentry was the dwelling. One particular log folk house became both diagnostic of the Upland South and an icon of the regional culture. Chapter 3 is devoted to it.

Chapter 3

The Upland Southern Folk House

Folk dwellings reveal much about their builders, inhabitants, and region. They speak of antecedents, diffusion, conservatism, lifestyles, habitat, and selection. The people of the Upland South made do with a relatively few log house plans—fewer than ten if you wish to "lump" rather than "split" types. One of these would become a regional icon.

Mythic Abodes

A mythic quality accompanies the American log house. To be born in one formerly boosted a politician's chance of becoming president, and even standing empty and derelict a log cabin causes passersby to stop and gaze. It embodies a national past at once romantic, genuine, heroic, and imagined (Figure 3.1). As Beulah Price wrote, "since pioneer days the log cabin . . . has come to be associated with God-fearing persons of honesty, integrity, and perseverance."[1]

In the Upland South, tens of thousands of log houses survive in the cultural landscape, more than a few of which remain inhabited. Though long in decline, they endure, often enshrined as little museums of a pioneer past. The scholarly literature on upland southern log houses compiled by geographers, folklorists, historians, and anthropologists is superabundant.[2] House types have been identified and named, origins proposed and debated, geographical distributions charted, and diffusions traced. In the process, scholars discovered two basic plans:"single-pen" and "double-pen," using the folk word "pen" to mean a basic unit of four log walls notched together. These could be a single story in height, or instead elevated to one-and-a-half or two stories. Variation in

FIGURE 3.1 *An English-plan, single-pen log house, reputedly the first Euroamerican dwelling built in Wise County, North Texas. The Tennessean builder stands on the porch. (Source: Paddock 1906, I, 310.)*

the placement of the chimney added subtypes to this limited repertoire of plans.[3]

The simplest and smallest upland southern house consisted of one square log pen, or room, sixteen to eighteen feet on a side (Figures 1.7 and 3.1). It had side-facing gables, one of which bore the chimney, and a door in both the front and rear walls. Of English origin, this house type is variously referred to as the "English-plan" or "single-pen" cabin. In many parts of the Upland South, it is the most common log house type.[4]

In upland southern double-pen log houses, three basic floor plan options existed. The two rooms could be placed abutting one another, with chimneys and fireplaces on each gable wall, a door connecting the pens, and usually two front doors (Figure 3.2). Called a "Cumberland" house, this type is most common in the plateau section of the southern Appalachians. In some counties, the Cumberland is the most numerous double-pen house type.[5]

The other options involved leaving a space between the two pens, either remaining open as a breezeway, or "dogtrot," or filled with a double-fireplace and chimney, creating a "saddlebag" house (Figure 1.16). An early Texas settler of North Carolina birth mentioned pioneer "double cabins" with "either a wide passage between or a big double chimney."[6]

FIGURE 3.2 *A "Cumberland" double-pen log house, characterized by two abutting pens, each with a front door, and side-gable chimneys, in Davidson County, Middle Tennessee. (Photo by the author, 1978.)*

The Saddlebag House

As it happens, both the saddlebag and dogtrot house types possess a diagnostic quality that helps explain the origin and character of upland southern culture, as I hinted in Chapter 1. Featuring a massive masonry chimney placed between its two log pens that provides ventilation for fireplaces in each room, the saddlebag house displays a striking visual appeal and symmetry (Figure 3.3). Its name, derived from the vernacular speech of the Upland South, was prompted by its resemblance to the balanced double load borne by a pack horse.[7]

Oddly, in view of the sizable literature on upland southern folk houses, only two studies

FIGURE 3.3 *The massive central chimney, venting two fireplaces, of a "saddlebag" double-pen house in Bryan County, Oklahoma. The chimney fills the entire width of the space between the two pens. (Photo by the author, 1981.)*

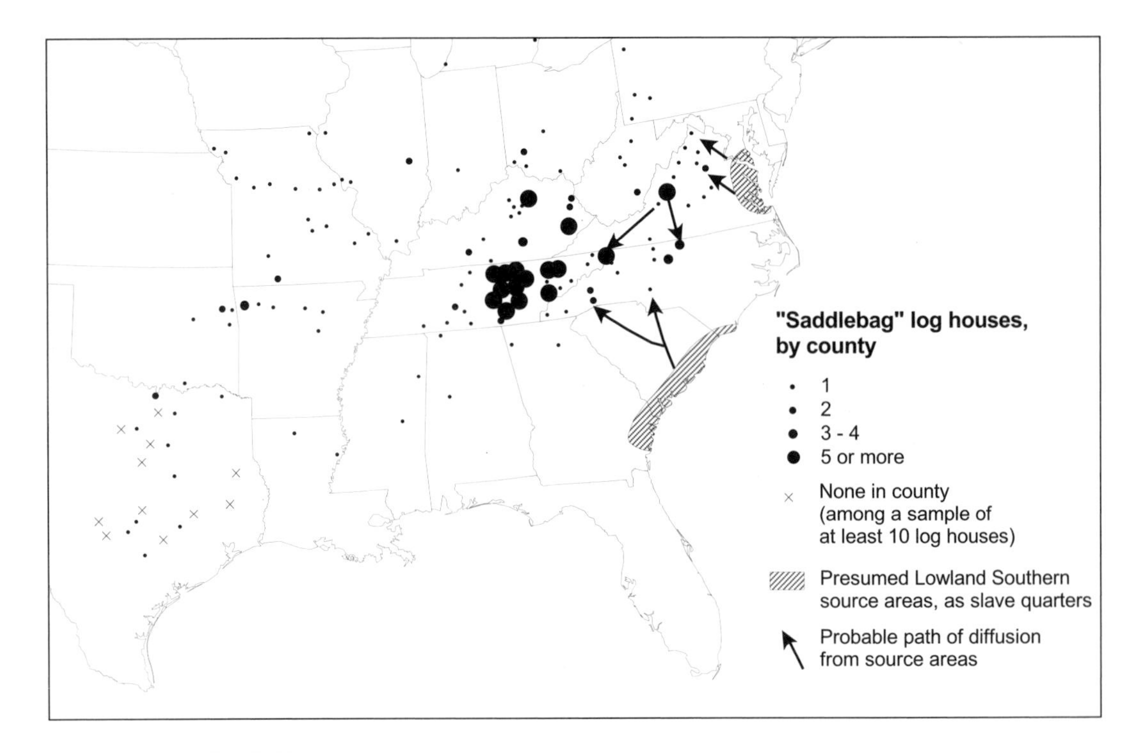

FIGURE 3.4 *Distribution of notched-log, saddlebag, double-pen houses in the South. This house type remained most common in the eastern and northern parts of the Upland South, where colder temperatures prevailed. It did not robustly accompany upland southern expansion. (Sources: Rehder 2002; Gavin 1995, 104–108; Morgan 1990; Dickinson 1990; Glassie 1965a, 161–162; Vlach 1972; and other sources listed in Jordan 1994, 40.)*

devoted exclusively to the saddlebag type exist.[8] Though it has received inadequate attention, the saddlebag still strongly suggests something important about the Upland South. Its origins are almost certainly British. In both Great Britain and Ireland, a prototypical saddlebag dwelling, built in masonry, provided servants' quarters on estates.[9] In that capacity, it found its way to the plantations of the Chesapeake Tidewater and South Carolina-Georgia Low Country as one type of slave cabin. By contrast, the British saddlebag house seems not to have become part of the Pennsylvanian colonial culture.

Carried inland by the planters, the saddlebag came astride the path of southwestward-migrating Midlanders, who quickly adopted it and rendered the plan in notched-log construction. This apparently happened fairly early in the southern back country, possibly in the Shenandoah Valley and almost certainly in the Piedmont (Figures 1.5 and 3.4).

As a result, the saddlebag house belonged to the Watauga tertiary nucleus of the Upland South from the very first. Indeed, Henry Glassie feels that a mature, symmetrical version of the log saddlebag arose precisely in the Watauga country.[10] Its link to that hearth is revealed both by the fact that the saddlebag achieved its greatest acceptance in central East Tennessee and adjacent southwestern North Carolina where it represents the most common double-pen house in multiple counties, and by the Missouri vernacular term "Tennessee house" to describe the saddlebag.[11]

Very rare in far northeastern Tennessee, the saddlebag accounts for only one among fifty log houses surveyed in Sullivan County and six

FIGURE 3.5 *A saddlebag, double-pen log house, Coles County, Illinois, known as the "Lincoln Cabin." From a postcard copyrighted by Dexter Press of West Nyack, New York, photo by C. L. Bence of Mattoon, Illinois. The successor, M. W. M. Dexter, Inc. of Aurora, Missouri, did not renew the copyright nor did Bence or his survivors.*

of 231 structures in adjacent Johnson County.[12] This suggests that its main route southward went by way of the Virginia Piedmont to the western North Carolina secondary hearth area, then through the Great Smoky Mountains into central East Tennessee. The saddlebag is occasionally described as a "Virginia type," revealing its Tidewater, Shenandoah, and northern Piedmont roots. It is also perhaps significant that the oldest known saddlebag house in Texas, dating to 1824 and still standing in the village of Independence in the old Austin's Colony where the Anglo-American settlement of Texas began, was built by a 1793 native of Rowan County, North Carolina, on the Piedmont.[13]

Oddly, though the log saddlebag diffused widely and very early through the Upland South, almost never did it achieve the status of most common house type or even double-pen (Figure 3.5). In Middle Tennessee, it is not all that common, though saddlebags appear with some frequency in the plateau area, surrounding the Nashville Basin.[14] Adjacent central Kentucky and Illinois likewise offer relatively few specimens.[15] Among 296 log dwellings in southern Indiana, only three saddlebags were found.[16] In spite of an early introduction of this house type by upland southerners in Texas, it remained rare. Nor is it common in Oklahoma or Louisiana.[17]

In short, the saddlebag log house reveals more about the origins of the Upland South than its geographical spread and extent. Viewing a map of its distribution reveals only a faint, almost ghostly trace of the Upland South at large (Figure 3.4). In part, the diffusionary lack of success of the saddlebag may have resulted from the heat-retaining quality of the central chimney, hardly an advantage in a subtropical climate. Also, the very people who would have best known and appreciated a central fire hearth—the Pennsylvania "Dutch," accustomed to their "continental" house—did not come in great numbers to the Upland South.

Instead, the companion double-pen log house—the "dogtrot"—provides a far better landscape indicator of the Upland South. The dogtrot would, in fact, become a regional icon.

Dogtrot Houses

The origin of the dogtrot house seems clear to me. A backwoods Finnish house type of the northern European forests, it reached the colony of New Sweden in the lower Delaware Valley in the middle 1600s.[18] A venerable Finnish-built dogtrot still stands near the banks of the Delaware River south of Philadelphia. It reached America as part of the log culture-complex imported from northern Europe that proved so influential in the Midland culture. Unlike the saddlebag house, the dogtrot first took root in North America in the Pennsylvanian culture, and most particularly in the pioneer phase, part of an outrageously successful backwoods forest colonization system (Figure 3.6).[19]

The dogtrot house plan represents the *simplest* way to build two pens or enlarge from single-pen size, an attribute that greatly enhanced the adaptive value. It is tedious and time-consuming to splice log walls, particularly if the first pen was constructed earlier, and a separate, second unit of four walls is far easier to add. If the second pen is situated eight to ten feet from the first, the skids normally employed to hoist logs into place in the upper wall can be employed. The point is that, if a structure of two log rooms was desired, the easiest solution was to space them apart in the manner of a dogtrot. Once introduced into the Midland pioneer architectural complex, it enjoyed a natural appeal and adaptive value based partly in ease of construction and enlargement.

In pioneer times, dogtrots normally resulted from adding a room to an existing single-pen house. Charles Martin's remarkable study of material culture in an upper valley of eastern Kentucky, where woodland pioneering persisted into the late nineteenth century, provides a documentation of house evolution in a frontier setting. One log dwelling recorded by Martin began as a single-pen in 1890 and was enlarged to become a dogtrot in 1905.[20]

An enlarged dwelling was normally needed in the frontier setting, since selection in woodland colonization favors a high birthrate and large families. The pioneer population is prolific because it is below average in age, can produce a large volume of food by exploiting abundant faunal and floral resources, and requires considerable labor for the

(top)
FIGURE 3.6 *A pioneer "dogtrot," double-pen log cabin, near Brownsville in far western Pennsylvania. Dogtrots were built in frontier times in Pennsylvania, but this house plan was never dominant there and almost none survive in the state today. (Source: Smith 1854, facing page 232.)*

(bottom)
FIGURE 3.7 *A double-crib log barn with open passageway, Denton County, North Texas. Its kinship to the dogtrot house is obvious, and the two—barn and house—spread together throughout the Upland South. This barn was razed in 1970, an altogether too-common occurrence. (Photo by the author, 1969.)*

continual task of clearing a forest. "Their little corn-patch increases to a field, their first shanty to a small log house, which, in turn, gives place to a double-cabin," noted an early observer on the Illinois frontier.[21]

Yet another adaptive advantage of the dogtrot plan was its remarkable versatility. Not only could it serve the needs of a large frontier family, but also a second room set apart from the first was well-suited to serve as an inn, summer kitchen, tavern, office, classroom, chapel, or jury quarters. The breezeway was not only a fine place to sit or sleep in warm weather, but also a convenient tackroom. When no longer needed as a dwelling, following the construction of a second-generation home, the dogtrot cabin easily became a barn. The breezeway then served as a wagon runway. Such "double-crib" barns, featuring an open-air passage between two log cribs, remain common in the Upland South (Figure 3.7).[22]

The dogtrot plan, then, enjoyed diverse adaptive advantages, not all of which were shared by the saddlebag. Add to these its suitability to a subtropical climate, where a breezeway became a true blessing. South of the border between the harsher continental climate of the north and the subtropical climate of the southeastern United States is precisely where the dogtrot house achieved its dominance. It enjoyed, in short, the "simple and strong statements of a functional structure."[23]

Dogtrot Diffusion

Spreading with the frontier, the dogtrot house went down the ridge-and-valley section of the Appalachians, though the diffusionary trace has gone faint (Figure 3.8). It spilled out onto the Piedmont, as evidenced by abundant examples in North Carolina, and also reached East Tennessee by the Appalachian Valley route (Figure 3.9).

One of the most rewarding moments of field research I have ever enjoyed occurred in the spring of 1984, when I happened upon a still-inhabited, aged dogtrot log house in the hamlet of Jordan Mines, Allegheny County, Virginia, situated in one of the numerous parallel valleys of the eastern Appalachians.[24] Its very existence revealed the Ridge-and-Valley corridor to have played a role in the diffusion of the dogtrot, and that discovery I found exciting. The present rarity of this house in western Virginia, along the paths of migration from Pennsylvania, as well as its almost complete disappearance from Pennsylvania reveal something important about the Upland South. The dogtrot house survived into the post-frontier era *only* in the Upland South, vanishing almost without a trace elsewhere in the Midland culture realm. Moreover, the dogtrot apparently never had been as common in

FIGURE 3.8 *A dogtrot house in Giles County, in the ridge-and-valley country of western Virginia, a Pennsylvanian migration route southwestward toward Tennessee. Chimney placement in the nearer pen suggests that the floor plan of the dogtrot house was still undergoing experimentation in 1783, when this house was erected. The house no longer exists. (Source: Johnston 1906, facing p. 398.)*

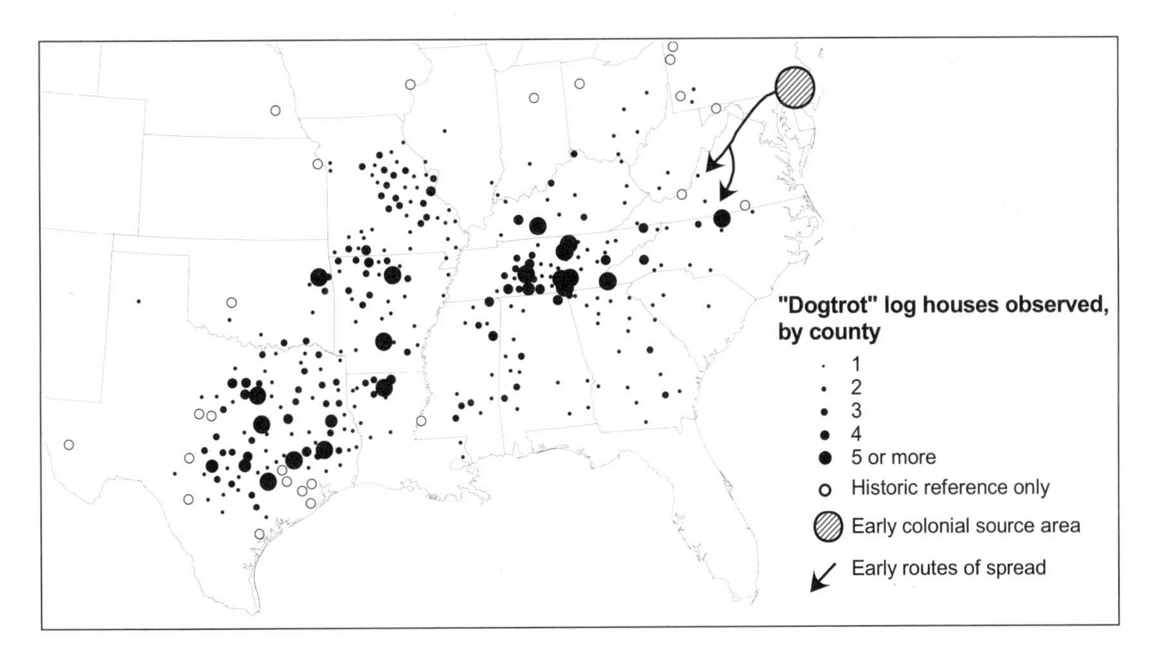

FIGURE 3.9 *Distribution of dogtrot log houses in the South. The pattern is spotty and incomplete, due to reliance upon field observations, but the spatial image conveyed is basically accurate. Many of these surviving structures have undergone modifications, especially the enclosure of the central passageway. Of all house plans, this one best epitomizes the Upland South. (Sources: Gavin 1995, 104-108; Rehder 2002; Wright 1956, 34, 46; Martin 1989, 140; and sources listed in Jordan and Kaups 1987, 56–57; and Jordan and Kaups 1989, 186–187; see, also, the sources for Figure 3.4.)*

Pennsylvania as it would become throughout the Upland South. It represents, then, both the magnification in importance of an element of Pennsylvanian culture *and* an archaism.

In western North Carolina and East Tennessee, the dogtrot achieved its first widespread acceptance and staying power, so much so that some scholars have mistaken this area as the place where the dogtrot originated.[25] Actually, the saddlebag house occurs more commonly throughout most of East Tennessee and adjacent North Carolina, yielding only in the southern part of the Valley of East Tennessee and far southwestern North Carolina to the dogtrot's dominance.[26]

Middle Tennessee perhaps better deserves the claim as the first area within the Upland South where ordinary people truly embraced the dogtrot as their regional signature folk house. More than one scholar has linked Middle Tennessee with the dogtrot.[27] Again, as with the half-dovetail notch, the role of the quaternary hearth of the Upland South in Middle Tennessee becomes clear. (Compare figures 1.5, 2.8, and 3.9.)

From Tennessee, and particularly from the middle part of that state, dogtrots spewed out to every corner of the Upland South (Figure 3.9). And with it, to almost every destination, went its distinction as the dominant house type.[28] Lestar Martin found 193 dogtrots, including both log and frame construction, in the hilly interfluvial parishes of northern Louisiana, and Texans embraced the dogtrot so fervently as to call it the "East Texas house."[29]

Along the northern margins of the Upland South, the dogtrot failed to achieve dominance, perhaps for climatic reasons. Warren Roberts found only seven such houses among 296 log dwellings in southern

Indiana, but then few Tennesseans went there.[30] Still, even in these northern margins, dogtrots did gain a foothold. One particularly fine example stands still today on a low bluff overlooking the east bank of the Mississippi River near its confluence with the Illinois (Figure 3.10). Nor did the Cherokees of Georgia and Oklahoma accept the dogtrot enthusiastically. Only nine percent of 333 log houses surveyed in the old Cherokee lands of northern Georgia are dogtrots, comparable to the nine specimens found in northeastern Oklahoma.[31]

Regional Icon

In addition to the mythic character enjoyed by log cabins in general, the dogtrot, by virtue of its dominance in Tennessee Extended, became a cultural icon for most of the Upland South. Wrote William Ferris, "the dogtrot has etched the memory and imagination" of upland southerners, becoming "a mythic image."[32]

Southern writers, painters, photographers, and architects found themselves drawn to the dogtrot.[33] William Faulkner used it repeatedly in his stories, as did Eudora Welty.[34] In time, as traditional upland southern culture weakened before the onslaught of American popular culture, the dogtrot became most often associated with poor whites.[35]

Surely much of the appeal of the dogtrot house lay in the breezeway. In Koeppen's humid subtropical climate, the "trot" replaced the primordial fireplace as the gathering place for the family (Figure 3.11). Here, after the day's work, the people assembled, seeking a cooling breeze while conversing, listening to stories, or singing. Many stayed on to sleep in the hot months. One Arkansas old-timer remembered the trot as the place "where melons were cut and horses swapped," while listening to "the music of pigs, poultry, and piano."[36] A European traveler in mid-nineteenth-century Texas described the open passage as being "according to the custom of the country" and offering its inhabitants "a cool, pleasant resort in summer."[37] Another Texas traveler of that time, turned away when he requested overnight accommodations during a rain storm, wrote that he "dragged my saddle under the covered corridor and slept."[38] From these functions, experiences, impressions, gatherings, and memories derived both the popularity of the house type and its iconographic status. In the brief interval between the advent of screened wire and air conditioning, the dogtrot enjoyed its finest hour.

Accompanying the dogtrot down from Pennsylvania, through Virginia, and into the Tennessee hearth was the double-crib log barn with open wagon runway (Figure 3.12).[39] From there the barn and house spread together through the remainder of the Upland South, companions always.[40] Lacking the dogtrot's function as a place to gather, the

FIGURE 3.10 *A magnificent two-story dogtrot house, beautifully situated above the eastern bank of the Mississippi River in Calhoun County, Illinois. (Photo by the author, 1989.)*

FIGURE 3.11 *A stuffed chair still sits in the breezeway of a recently abandoned dogtrot house in Panola County, East Texas. The "trot" replaced the ancient fireplace for family gatherings and for receiving guests during much of the year in the Upland South, helping make the dogtrot house a regional icon. (Photo by the author, 1981.)*

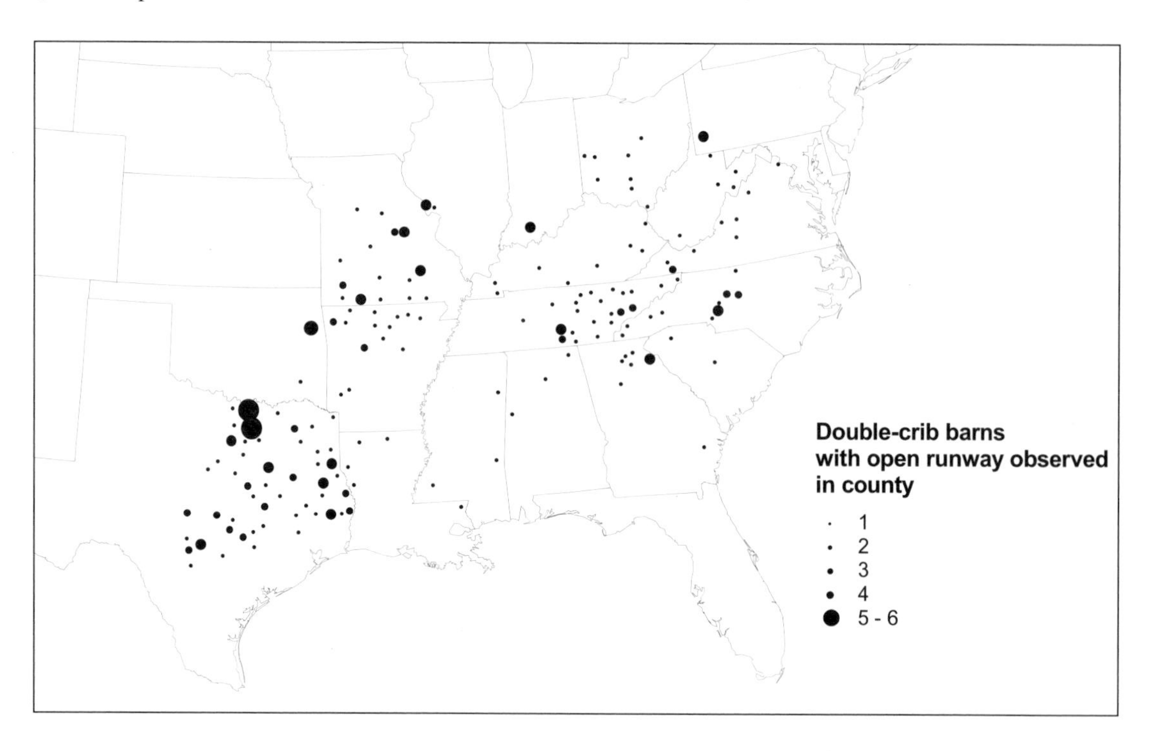

FIGURE 3.12 *Distribution of log double-crib barns with an open passageway observed in field research in the American South. Compare the pattern to that of the dogtrot house, shown in Figure 3.9. (Sources: Clendenen 1973, 117–119; Riedl et al. 1976, 214–217; Wright 1956, 94, 99; Glassie 1965–66, 12; Stuck 1971, 228, 231, 233; Rehder 2002; Milbauer 1996–97; and Jordan and Kaups 1989, 190-191.)*

double-crib barn would never become a regional icon. Poets do not pursue it. In fact, the double-crib, though very common, did not even become the dominant barn type of the Upland South. That honor would belong to the transverse-crib barn, the subject of Chapter 4.

Chapter 4
Inventing a Barn

UPLAND SOUTHERNERS devised a system of highland farming even more distinctive than their dwellings or carpentry. Oddly, no one has ever given it a name, so let's call it "upland southern mixed farming." Geographically the system has pretty clear boundaries (Figure 4.1). To the south lies the old coastal plain realm of the cotton plantation system, now given way to neoplantations raising more pine trees and soybeans than cotton. Westward live cattle ranchers and cash wheat farmers, and to the north the upland southern agro-economic system borders the Corn Belt and Dairy Belt.[1]

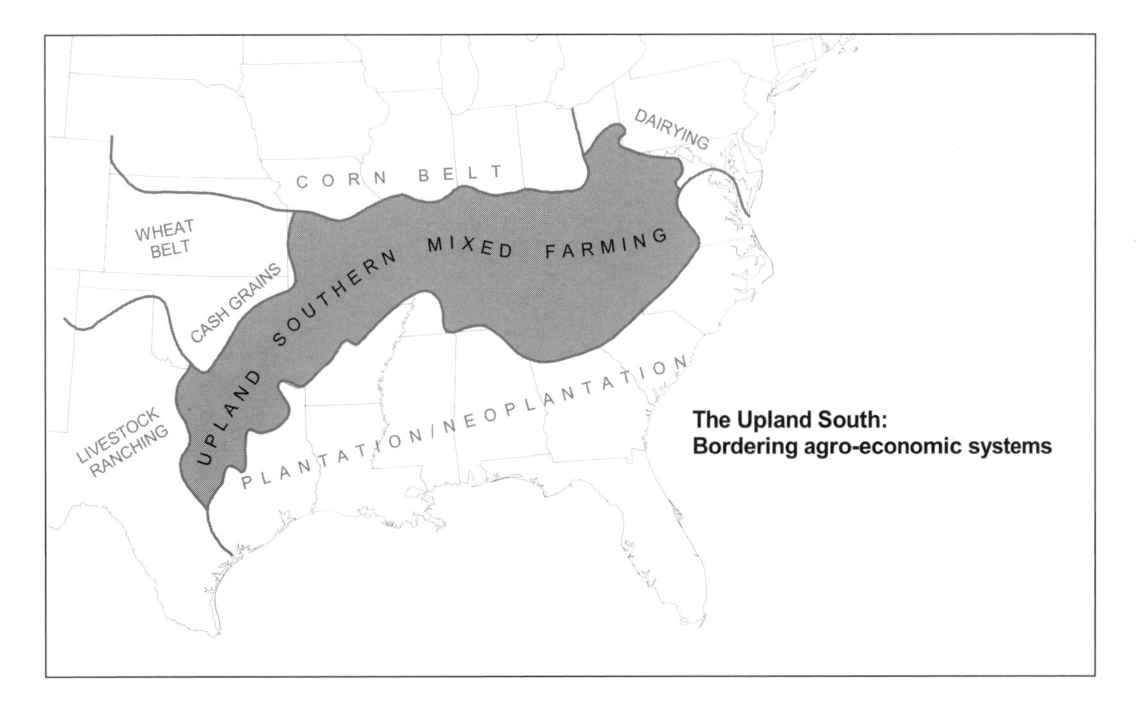

FIGURE 4.1 *Agro-economic systems of the South and bordering areas. (Sources: Baker 1927; Hudson 2000, 76–77.)*

Upland Southern Mixed Farming

Traditional farming in the Upland South possessed both diversity and a high degree of self-sufficiency, but by no means was it an isolated subsistence system. "Independent but not isolated" is the way one scholar described it.[2] This agricultural operation included small numbers of open-range cattle, droves of hogs and flocks of turkeys in the woods, a small herd of sheep, a horse and milk cow or two, maize fields, patches of oats and tobacco, hay cut from unimproved meadows to bring the unsheltered stocker cattle through the winter, and a sizable kitchen garden/orchard. From this multifaceted operation, the upland southern farm family, properly called "yeomen" or "plain folk," supplied most of their needs and sent to market lean cattle, pigs, free-range turkeys, corn whisky, and perhaps a bit of tobacco and a few mules.[3] By marked contrast, Pennsylvania's other, more ambitious child—the lower Midwest—acquired the highly specialized and commercialized German-inspired stock-fattening system we eventually came to call the Corn Belt.

Maize grew in a land rotation system, or "forest fallowing," in which fields were abandoned after some years of production and a new clearing made, allowing fertility to be restored in a decade or two before the same land was cleared again.[4] The corn patch was "fenced in," leaving the bulk of the land as open range, and, when the farmer abandoned a field, the "stake and rider" rail fence, which had no posts, was typically disassembled and taken to the new clearing (Figure 4.2).[5]

FIGURE 4.2 *A typical upland southern corn field, bordered by a stake-and-rider "worm" fence, near the Hill Country hamlet of Bee Cave, Travis County, Texas. More recently, this place has undergone suburbanization in the sprawl of the city of Austin, but the fence survives, now as the border of a parking lot. (Photo by the author, 1968.)*

FIGURE 4.3 *A log corncrib of single-crib size, a typical outbuilding throughout the Upland South, in Denton County, North Texas. An undersided saddle notch was employed by the builder. (Photo by the author, 1973.)*

Corn, without serious rival, was the most important crop, providing meal, a bit of feed, and whisky. In one form or another, it provided much of the farmers' diet and commercial production by value (Figure 4.3).[6] Near the farmhouse lay a sizable garden and orchard. In addition to an array of fruit and vegetables, a small patch of tobacco often grew there. Close by stood chicken coops, milking pens, and a sty for pigs being fattened for slaughter.

The free-range countryside served both as hunting grounds and housed the cattle, hogs, and turkeys, marketed lean by middlemen drovers using established trails leading to destinations outside the Upland South.[7] The countryside also provided fishing streams and a place to gather herbs, berries, and nuts.[8]

To a considerable degree, the upland southern mixed farming system was archaic, preserving elements of the Midland frontier backwoods colonization system that had emanated from Pennsylvania and the lower Delaware Valley.[9] It joins the list of items that explain the Upland South as a regional repository of archaic Pennsylvanian culture. But hill farming became more than that. It evolved into a post-pioneer stability, becoming somewhat more market-oriented in the process. In time, it degenerated into rural poverty, more from overpopulation than anything else, allowing mining, logging, welfare, and tourism to play larger and destructive roles. Eventually the farming system disappeared from the countryside, perhaps about 1950 or 1960, but many vestiges remain.

The Farmstead

In the pioneer era, the upland southern farmstead had contained few structures other than the dwelling. A small single-crib for maize storage sufficed as a granary (Figure 4.3), or perhaps a double-crib barn at most.

As the upland southern mixed farming system matured in post-frontier times, the number of buildings proliferated, eventually forming a sizable farmstead assemblage.[10] A springhouse, smithy, smokehouse, chicken coop, sty, and an array of other small specialty structures appeared. But most of all, the successful, middle-class upland southern yeomen required a more sizable barn, one possessing greater capacity than the double-crib. They had to invent one or, as it happened, substantially modify and enlarge a type already known to them. It would become known as the *transverse-crib* barn, so closely linked to the regional culture as to be labeled, only half-facetiously, as the "ethnic barn of the Upland South (Figure 4.4)."[11] While it would never become a regional icon like the dogtrot house, the transverse-crib barn would eventually be adorned with the lion's share of the ubiquitous "Mailpouch" chewing tobacco and "Rock City" painted wall advertisements.

Defining the Transverse-Crib

The multifunctional transverse-crib barn efficiently served many of the needs of the diversified mixed-farming system of the Upland

FIGURE 4.4 *A frame transverse-crib barn with a cupola atop the roof, vertical siding, and one eave side shed, in Monroe County, part of Missouri's "Little Dixie." The type is here near the northern edge of its geographical area of occurrence. (Photo by the author, 1990.)*

South. But defining it is not simple, because classifications differ according to which traits receive emphasis and also because transitional types invariably occur. Add "lumpers" and "splitters" to the debate, and the likelihood of consensus wanes further. I choose to lump. My criteria for a transverse-crib barn include: (1) gables facing front and rear; (2) a central through-passage runway directly beneath the roof ridge and having wagon access at both ends; (3) four to ten cribs (most typically six) situated on either side of the runway; (4) a loft positioned above the cribs; and (5) multipurpose functions, including at a minimum a threefold division among granaries, stalls for a few draft animals and/or milk cows, and hay storage (Figure 4.5). A versatile structure, the transverse-crib may also have cribs devoted to gear rooms or

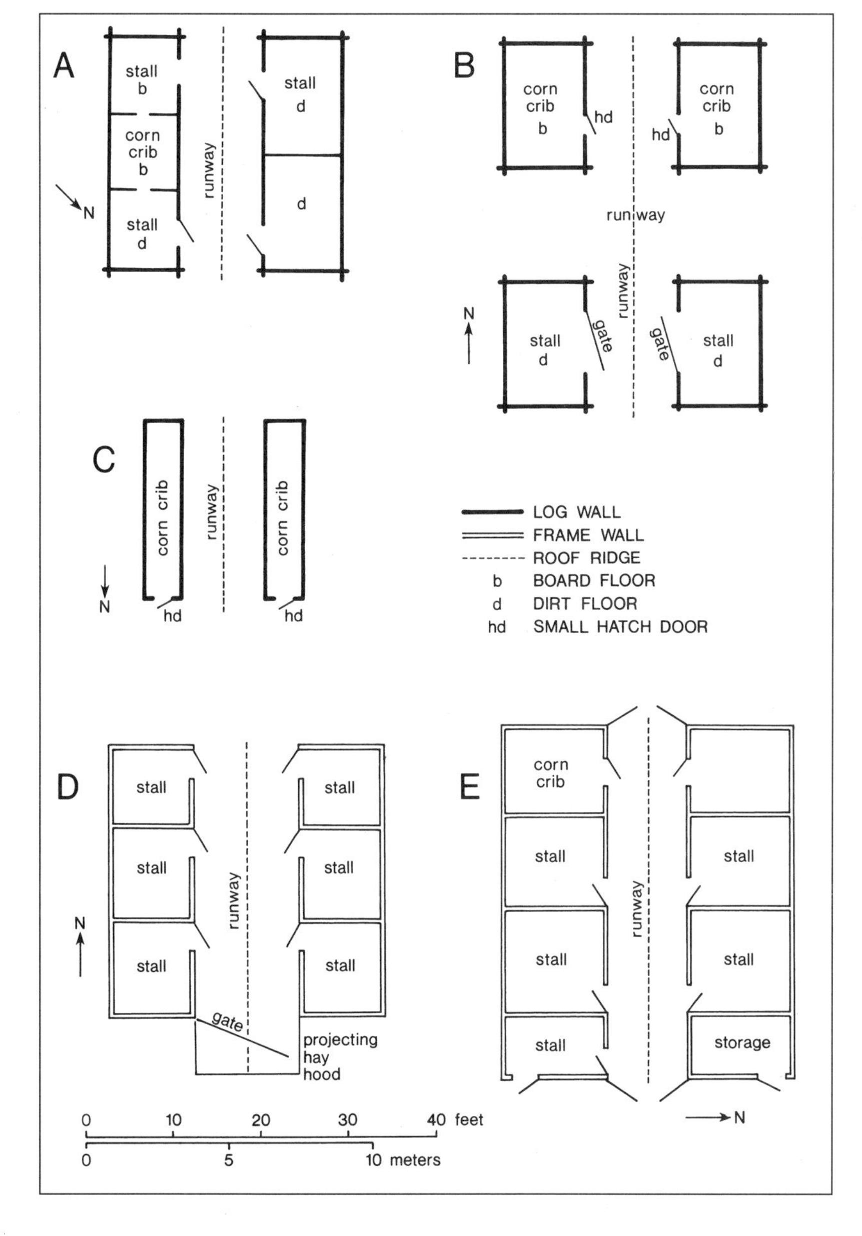

FIGURE 4.5 *Plans of representative barns: A = transverse-crib near Paintsville, Johnson County, Kentucky, built of saddle-notched, round poplar logs; B = four-crib barn near Flower Mound, Denton County, Texas, ca. 1880, with elongated cribs made of saddle- and-V-notched post oak logs; C = drive-in corncrib at Brattonsville, York County, South Carolina, consisting of hewn, half-dovetailed logs; D = frame transverse-crib, Hardin County, Illinois; and E = frame transverse-crib, ca. 1850, near Yates, Howard County, Missouri. All shed additions and frame additions to log barns have been deleted from the plans, to emphasize the basic barn. (Sources: Marshall 1981, 73; Sculle and Price 1993, 18; Jordan-Bychkov 1998, 8.)*

FIGURE 4.6 *Horses seek scattered hay remnants in the open runway of an Independence County, Arkansas, transverse-crib barn. (Photo by the author, 1989.)*

diverse other types of storage, and grains can be husked, shelled, or threshed in the runway (Figure 4.6). Sometimes a small harvest of tobacco is cured in side shed additions.[12] But the basic threefold combination of granary/haymow/stall defines the type functionally. To meet my definition, then, the transverse-crib must be more than a mere granary with haylofts, and for that reason the so-called "drive-in corncrib" of the Appalachians and Midwest, while similar in form though smaller, cannot be regarded as a transverse-crib barn (Figure 4.7).[13]

FIGURE 4.7 *A log "drive-in corncrib," York County, South Carolina. This sort of outbuilding, which occurs fairly widely in the eastern part of the Upland South, may be genetically related to the transverse-crib, but it is far smaller, serves fewer functions, and consists of only two, elongated, narrow cribs. (Photo by the author, 1996.)*

Negative definitions also prove useful. The transverse-crib barn almost never has a banked wagon ramp access to the loft level, is never used exclusively for tobacco curing, and does *not* function as a feeder barn for producing fattened livestock. Efforts to trace Corn Belt feeder barns to the Upland South are singularly misguided, since these hill folk never engaged in the feeder business.[14] Nor is the transverse-crib to be confused with outwardly very similar types such as the "Dutch" barn of New York and New Jersey; the tobacco sheds of the Kentucky Bluegrass Country, which lack internal crib division; the "Cajun" barn of Louisiana; certain Midwestern feeder barns; and the Rocky Mountain horse barn of the West.[15]

Variant Forms

Since my definition of the transverse-crib barn is in most respects inclusive, some major variant forms need to be acknowledged. For one thing, material composition differs. The older transverse-cribs are built of notched logs, using the venerable upland southern carpentry techniques, though relatively few log examples survive (Figure 4.8). The transverse-crib barn represents a post-frontier development. Most were built after notched logs had given way to frame construction. It is not Daniel Boone stuff, but instead reflects the need for a larger barn that arose when the backwoods colonization system gave way to the post-pioneer, more intensive land-use strategy.

FIGURE 4.8 *A log transverse-crib barn, in Maury County, Middle Tennessee. This is, I feel, the prototypical form (minus the side shed). (Photo by the author, 1993.)*

FIGURE 4.9 *A log four-crib barn, Taney County, Missouri Ozarks. (Photo courtesy State of Missouri, Dept. of Natural Resources, Division of Parks & Historic Preservation, Neg. no. 45, kindly provided by James M. Denny, Chief, Survey & Registration.)*

Most often, log specimens of the transverse-crib consist of four cribs, though some contain as many as eight, perhaps most notably the one preserved as a part of the "pioneer farmstead" at Oconaluftee, in that section of the Great Smoky Mountains National Park lying in Swain County, North Carolina. This remarkable barn, much trifled with by the National Park Service, is covered by a huge, detached canopy roof. Two main subtypes of log transverse-cribs occur (Figure 4.5, A and B). One consists of four freestanding cribs bound together by an overarching roof. Two wagon runways form a cross between the cribs. Some Texas farmers call this a "foursquare" barn, perhaps a venerable upland southern term now nearing extinction, but barn scholars call it a "four-crib" (Figure 4.9).[16] In the second log subtype, the eave-to-eave passage is absent, so that cribs abut one another on each side of the single remaining runway (Figure 4.8). Henry Glassie called this a "log transverse-crib" to distinguish it from the double-aisled foursquare type.[17] A splendid barn of this type stands in Orme, Marion County, Tennessee, right on the Alabama state line. Owned by the Phillips family, it measures fifty-three by thirty-two feet, with multiple plate logs of white poplar that span the entire dimensions. Half-dovetailed, it consists of eight cribs but is highly assymmmetrical, with five cribs on one side, each about ten feet wide, and three on the other, one of which has a doorway passage leading out to the eave side. Such assymetries suggest a continual process of experimentation and modification of the log transverse-crib barn and also reveal its versatility of form.

FIGURE 4.10 *A small frame transverse-crib barn, Randolph County, Alabama, consisting of only four cribs and a rather confined loft. This represents the smallest type of transverse-crib. Note the horizontal siding and unbroken roof pitches. (Photo by the author, 1988.)*

The overwhelming majority of extant transverse-cribs, well over 95 percent, are built of frame construction, with milled boards affixed either vertically or horizontally as siding (Figures 4.4 and 4.10). The dominance of frame construction reveals the success of the transverse-crib in facilitating the transition from the pioneer folk phase of upland southern culture to the twentieth century. Frame transverse-cribs were still being built in some parts of the Upland South as recently as 1950.

Size also varies considerably. The enduring popularity of the transverse-crib in no small part results from its versatility in dimensions. Cribs can number from four to six, eight, ten, or conceivably even more, though I have never seen one larger than ten cribs.[18] Variation in the number of cribs allows an impressive lateral size range. Nor did this gable-to-gable enlargement exhaust the possibilities for lateral expansion. Shed additions on one or both eave sides, either enclosed by board walls or open-sided, are ubiquitous (Figure 4.11). The farmer might choose, instead, to employ one or both side sheds as additional passages, parallel to the central wagon runway and often used for sheltering farm machinery or curing tobacco. If two such side passages exist, the term *three-portal* barn is used to describe the enlarged transverse-crib, since the structure has triple entries on each gable end.[19] An ultimate lateral expansion, which at last exhausts such possibilities, can be achieved by adding a row of cribs beyond each of the two side passages, producing a formidable barn with four crib rows and three runways (Figure 4.12).

FIGURE 4.11 *A large frame transverse-crib barn, Harrison County, West Virginia. The haylofts extend out above the sheds on the eave sides. (Photo by the author, 1991.)*

FIGURE 4.12 *A frame transverse-crib barn of the "three-portal" subtype, Lewis County, West Virginia. Shed storage space has been built beyond the two side passages, creating a quite spacious barn. (Photo by the author 1991.)*

FIGURE 4.13 *A frame transverse-crib with a "hay hood" and doors to close the runway, Smith County, Middle Tennessee. The barn is painted blue. (Photo by the author, 1993.)*

Lateral rather than vertical expansion typified the houses and barns of the Celtic peoples of the British Isles, including the Scotch-Irish. Germans in Europe habitually did precisely the opposite, enlarging upward. We should not be surprised, then, that a barn type possessing a major capacity for lateral expansion should become dominant in the Upland South, a culture region in which Scotch-Irish heritage plays a large role.[20]

This should not imply that vertical enlargement of the transverse-crib barn is impossible. Some, in fact, attain an impressive height (Figure 4.12). Enlargement, however, differs from expansion. Making transverse-cribs taller invariably serves only to enlarge the hayloft, making it more than a mere attic. As a result, the transverse-crib never becomes a proper multi-level structure, unlike the German-derived Pennsylvania forebay bank barn, which has an upper floor accessible by wagon.

Other variations displayed by transverse-crib barns are less consequential. The wagon runway may be enclosed by doors, left completely open as a breezeway, or equipped with slat fence gates. A "hay hood" or "hay bonnet," containing a lifting device, may jut from the upper part of the front gable, but usually hay was simply forked up from the wagon into the loft (Figure 4.13).[21] Roof profiles vary from the gambrel, a newer type which enlarges the loft space, to the more traditional type consisting of two unbroken pitches (Figure 4.14). Shed additions often have roofs of a gentler slope than the main barn structure, but many sheds are tucked beneath roofs of unbroken profile (Figures 4.11 and 4.12). "Monitor" roofs, featuring a raised ventilator for the loft at the center of the roof ridge, occur occasionally on transverse-crib barns, and cupolas and other smaller ventilators are common (Figure 4.4). Finally, color also varies. I have seen red, white, blue, and even black transverse-cribs, but the majority remains unpainted.

FIGURE 4.14
A gambrel-roofed, frame transverse-crib barn, in Washington County, Missouri Ozarks. (Photo by the author, 1989.)

Geographical Distribution

The transverse-crib barn is exclusively and uniquely an upland southern type (Figure 4.15). No students of the American barn seem to disagree on this point. Fred Kniffen found it "throughout the Upland South and periphery" as "the dominant barn," words echoed by Doug Meyer, who also judged the frame transverse-crib to be dominant in the Shawnee Hills of southern Illinois.[22] In neighboring Indiana, Robert Bastian concluded that the transverse-crib was "common to prevailing" in the southern part of the state, particularly the "hill country of south-central-Indiana" and excluding only counties fronting the Ohio River.[23] Glassie said transverse-cribs occur "throughout the southern Appalachians."[24] In Texas, the type remains fairly common, and even as far afield as eastern Kansas the transverse-crib is labeled a "southern" barn.[25]

I would go a step further and suggest that the transverse-crib barn is *diagnostic* of the Upland South, both because it reflects the diversified mixed-farming system underlying that subculture and because it reveals the Celtic preadaptation and preference for lateral rather than vertical expansion of structures. When you encounter assemblages of these barns, you know you are in the Upland South, just as the sight of bank barns conveys the visual message that you have departed that region.

Even so, the distribution of the transverse-crib barn within the Upland South is far from uniform. Words such as "throughout" and "dominant," tossed out so glibly, do not hold up under close scrutiny. A zone of transverse-crib dominance occurs in Middle and much of East Tennessee, lapping over into rather narrow peripheries of all bordering states, as well as into western South Carolina. Coffee County, in Middle Tennessee, where in-depth field research detected fifty-one transverse-

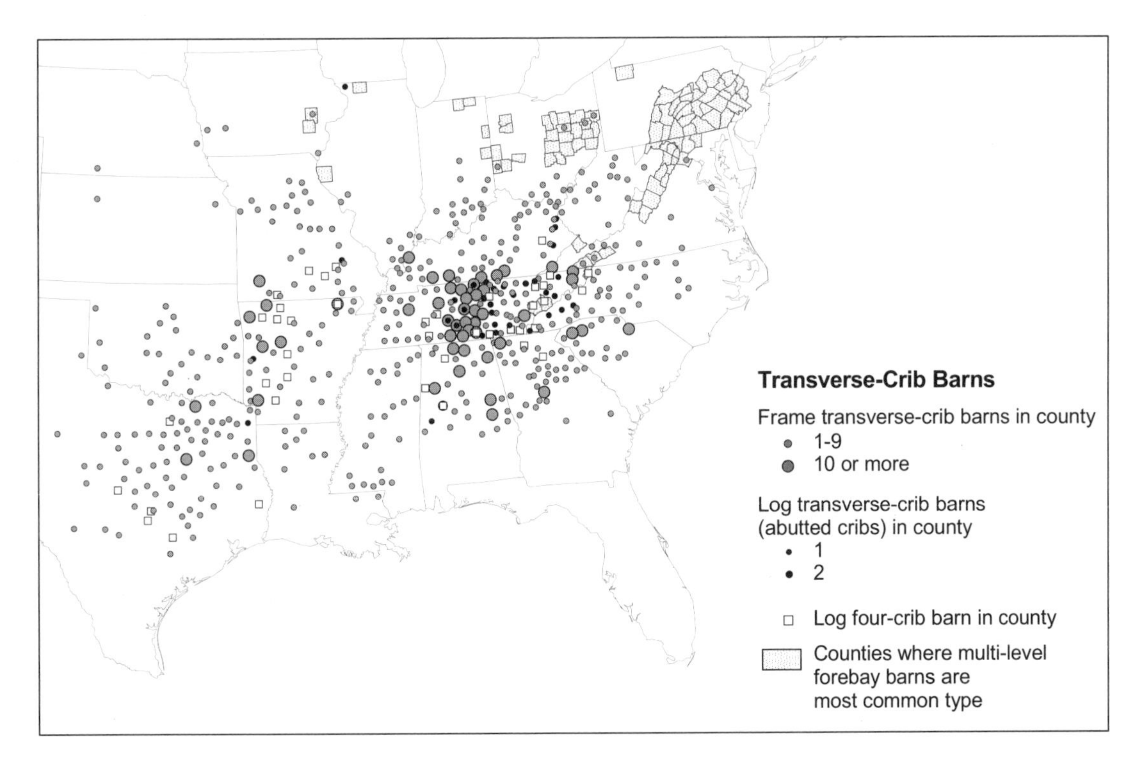

FIGURE 4.15 *Distribution of transverse-crib barns. The map is based largely upon my own field observations, made during the period 1970 to 2002, but I have annexed data, both published and unpublished, from many other scholars and barn buffs. They are listed in Jordan-Bychkov 1998, 12, 26–32. (Other sources: Rehder 2002; Glassie 1965–66, 18; Ensminger 1992; Glass 1986; Morgan 1997.)*

crib barns in one rather confined area, and nearby Rutherford County, where I observed thirty-three transverse-cribs in a cursory survey, epitomize this Tennessean core region, as do neighboring Smith County with twenty-seven and East Tennessee's Grainger County with twenty-five.[26] Secondary, lesser concentrations appear in parts of the Ozark and Ouachita mountains of Missouri and Arkansas, as well as in southern Illinois, where Keith Sculle and Wayne Price found 163 specimens in Hardin County alone.[27]

The only serious challenge to the transverse-crib in a competition to provide a larger barn for the upland southern mixed farming system came in southwestern Virginia and the Watauga country. There, Pennsylvania-German-inspired barn builders placed a massive second level atop log double-crib and four-crib barns, with visually striking overhanging forebays on two or all four sides.[28] This tall Teutonic type became the dominant barn in a handful of counties but failed to gain a wider acceptance among upland southern farmers, probably due to the weakness of a German ethnic presence south of the Shenandoah Valley. But simply the fact that the Watauga country witnessed this contest between the transverse-crib and "Dutch" forebay barns suggests that we look to the quaternary hearth in Middle Tennessee as the district where, at last, the ascendance of the transverse-crib barn reached its culmination and the Upland South coalesced.

The rarity or absence of transverse-crib barns in certain areas tends to negate an upland southern identity. This is certainly true of the "Delta" country along the lower Mississippi River valley, which drives a wedge northward almost severing the Upland South. The relative absence of transverse-cribs in the Delta, as well as in the outer Gulf coastal plain and below the Fall Line in the south Atlantic states, underscores the unsuitability of this barn type to the traditional plantation system. Similarly, the transverse-crib had no role to play in the cattle ranching economy of the Great Plains and failed, for that reason, to accompany the upland southerners who settled West Texas. Many counties in that part of Texas have Tennessee-derived populations but no transverse-crib barns. Nor did the barn offer any utilitarian value to cash grain farmers. As a result, transverse-cribs did not penetrate the Wheat Belt.

Moreover, as earlier noted, the transverse-crib never became the barn of choice in the Corn Belt, yielding to German-inspired, multi-level bank barns in most of Ohio, Indiana, and Illinois. While you will see on many Corn Belt farms a small structure looking very much like a transverse-crib barn, closer inspection invariably reveals it to be merely a slat-sided drive-in corncrib, usually lacking even a hayloft. Nearby stands a much larger structure—the true Corn Belt feeder barn.

Surprisingly, several areas usually regarded as parts of the Upland South have few if any transverse-crib barns. If we are to believe the cultural diagnosticity of the barn type, then very little of Virginia can be regarded as upland southern. Large parts of Kentucky and West Virginia also fall away, and we would have to reject Wilbur Zelinsky's proposal that southwestern Pennsylvania might have upland southern affiliations.[29] In southern Indiana, too, the role of upland southern influence has perhaps been overstated, to judge from barn evidence.

European ethnic minorities who settled amongst and adjacent to upland southerners rejected the transverse-crib almost unanimously. This was true of the Germans, Scandinavians, and Czechs in Texas and of the Germans in Missouri as well. It remained a barn of the old-stock Anglo-Saxon element.

The geographical distribution of the transverse-crib barn reveals not only the territorial extent of the Upland South, but also a core/periphery configuration of the culture (Figure 4.15). If we can trust the diagnosticity of the transverse-crib as a visible index to both the presence and intensity of upland southern influence, then we should speak in no uncertain terms of a Middle Tennessee hearth and core area.

Transverse-Crib Origin

Such speculation demands an attempt to pinpoint more precisely the origin of the transverse-crib barn. We can, I feel, safely reject any notion of European origin. True, log barns very much like the transverse-crib exist in the Alps, and an Old World tradition of central-aisled, front gabled barns is widespread, even pervasive, in many areas of

Europe.[30] But where are the missing links—European barns of the American colonial seaboard that might have inspired the transverse-crib? Pennsylvania and the southern coastal plain lack these possible prototypes.

The front-gabled, central-aisled Dutch barn of New Netherland has been suggested as the precursor of the transverse-crib, but far too wide a territorial gap exists between the scattered New York/New Jersey pockets of Dutch barns and the upland southern heartland.[31] Nor is southern Appalachia noteworthy as a gathering ground for Dutch migrants from the Hudson Valley. Moreover, the New World Dutch barn, while outwardly similar to the transverse-crib, has a very different floor plan. Internal division into cribs normally does not exist, and the central aisle is usually flanked by open feeding stalls, mow, and semi-enclosed stables. A rear wagon entrance is often absent, and smaller doors as well as windows appear in locations not found in the transverse-crib.[32] The Dutch barn is a fundamentally different type.

No, the transverse-crib seems to represent that rarest of objects—an Americanism from the era of folk culture.[33] Almost every other material aspect of Pennsylvanian and upland southern culture has demonstrable European prototypes and clear paths of diffusion. The transverse-crib barn, by contrast, appears to be indigenous to the southern back country, and on that point we find general agreement.

Both Kniffen and Glassie believed that the transverse-crib evolved in the early 1800s when farmers, desiring a larger outbuilding, doubled the size of the open-runway, log two-crib barn, a European type that spread widely from southeastern Pennsylvania.[34] This cloning reputedly produced the four-crib barn, with its odd, crossed runways. As we saw with the dogtrot house, log cribs are most easily erected when free-standing, allowing the builders to use skids to raise the logs into place, and the advantage of the four-crib plan was that it consisted of free-standing units, tied together only by the massive roof. Farmers, realizing that only one of the crossed runways was needed, supposedly next blocked off the side-to-side passage by enclosing two additional cribs in its place, creating the true transverse-crib plan. The gable-to-gable runway best served the function of offloading hay into the loft, since it had more headroom, and for that reason it was the one to survive. Also, gabled entrances permitted the barn to be further enlarged laterally with additional cribs, an expansion less feasible on the eaves. Glassie found one odd six-crib log barn in northern Alabama with side gables, and his photograph of it clearly reveals several practical reasons why the front-gable plan prevailed.[35]

Kniffen and Glassie both felt that the most likely area of origin of the transverse-crib barn lay in the southern part of the Valley of East Tennessee, in the southeastern corner of the state.[36] In an early publication, Glassie suggested that certain four-crib barns of the Great

Smokies represented the oldest, prototypical type. These displayed asymmetries in plan, such as rectangular rather than square cribs, runways of unequal width, and inconsistent crib door placement. Later four-crib barns, he claimed, were symmetrical and square in plan.[37]

I disagree. In my field experience, which includes inspection of twenty-five or so four-crib barns, I find no geographical or temporal concentrations of the asymmetrical plan. Some fairly late four-crib log barns in Texas display most or all of the plan irregularities Glassie ascribes to the prototype (Figure 4.5). See, for example, the quite asymmetrical four-crib barn preserved at Common's Ford Park in Austin. More plausible, in my opinion, is the evolution of the transverse-crib from Glassie's "type III" or "Alpine" double-crib log barn, characterized by two elongated, subdivided cribs—in effect making four cribs out of two—and by side-facing gables (Figure 4.16). Its ground plan is identical to smaller log transverse-cribs (Figure 4.8). All one need do to create a transverse-crib from the type III is rotate the roof by 90°, a modification that brings the structural advantages described earlier. This double-crib subtype apparently came over from Alpine Europe to colonial Pennsylvania, then spread down the Great Valley of the Appalachians to East Tennessee.[38] While the initial transition from Alpine double-crib to transverse-crib involved a mere roof modification, the subsequent enlargements by upland southern barn builders made the transverse-crib more properly an *invention* by a process of incremental change.

Deriving the transverse-crib barn from the Alpine double-crib means that the four-crib, or foursquare, plan is the *child* rather than the parent of the abutted-crib type (Figure 4.17). To me, that makes far bet-

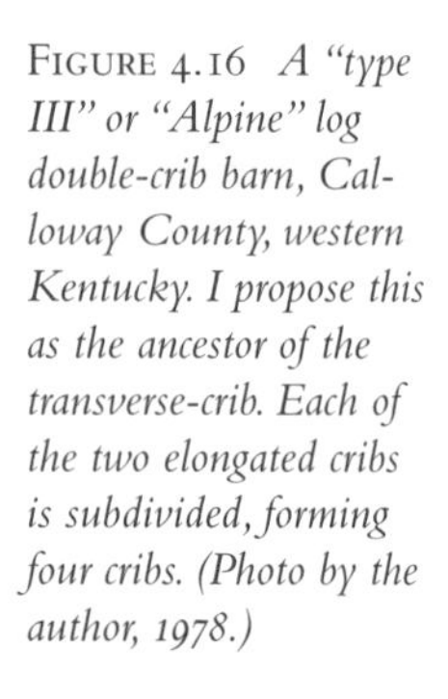

FIGURE 4.16 *A "type III" or "Alpine" log double-crib barn, Calloway County, western Kentucky. I propose this as the ancestor of the transverse-crib. Each of the two elongated cribs is subdivided, forming four cribs. (Photo by the author, 1978.)*

ter sense. Builders would have erected four-crib barns whenever they could not locate timbers long enough to span the lengthy eave side walls. The farmer would also have opted for the four-crib plan if a larger barn was desired, since immediate closure of the side-to-side passage with boards produced a six-crib structure. Both circumstances—the rarity of long timbers and the desire for six cribs—would more likely have occurred later in the settlement history, supporting my notion that the four-crib derives from the earlier log transverse-crib.

This scenario also allows us to make sense of two odd facts. First, the four-crib plan *never* appears in frame construction. Surely, if it were the parent form, that would not be true. Second, the four-crib barn appears in most areas where the frame transverse-crib occurs, as far afield as Central Texas. It displays an odd, strewn geographical distribution, with no hint of a concentration in some potential hearth area.

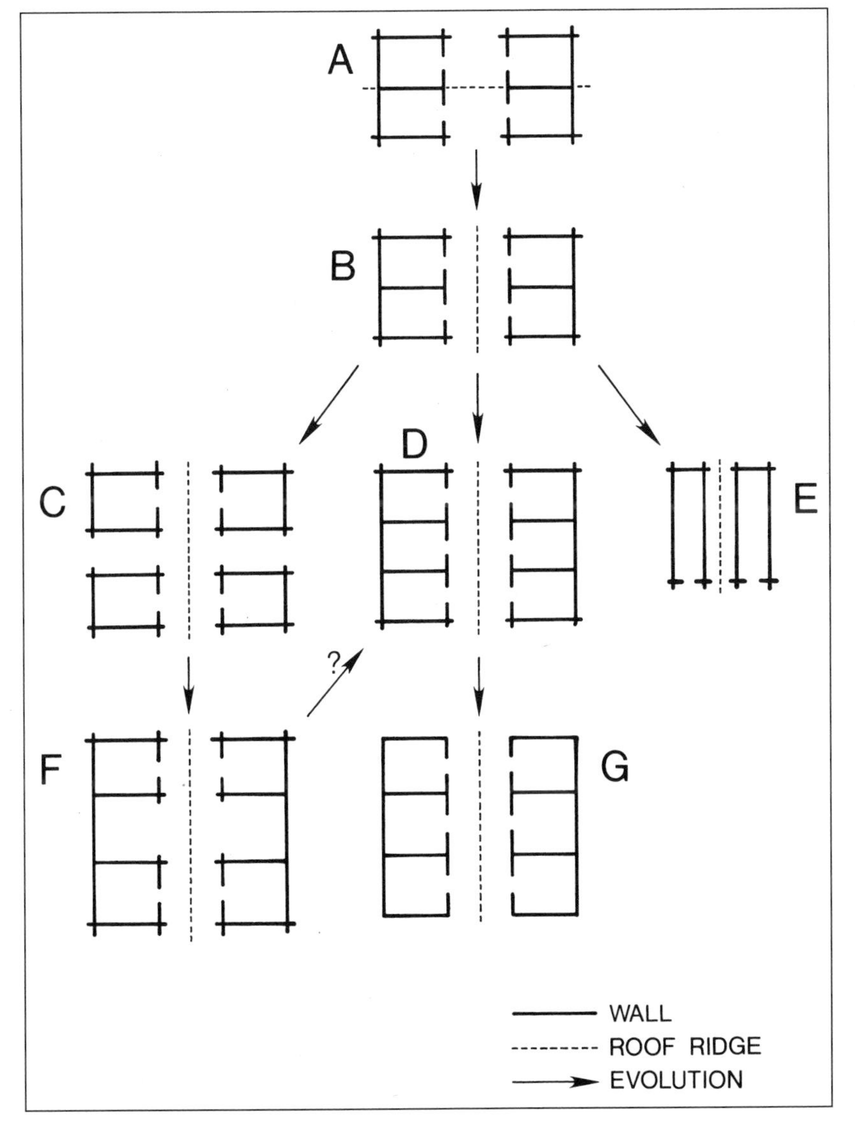

FIGURE 4.17 *Suggested evolution of the transverse-crib barn and related types. A = "Type III" or "Alpine" double-crib log barn, with subdivided cribs; B = prototypical log transverse-crib, consisting of four cribs; C = log four-crib barn; D = enlarged transverse-crib, with six log cribs; E = drive-in corncrib; F = log four-crib barn with one runway enclosed; and G = frame transverse-crib consisting of six cribs.*

FIGURE 4.18 *A fine, large transverse-crib barn at Fairview, Washington County, East Tennessee, in the heart of the old Watauga settlements, where this barn type likely originated. Note the atypical roof ventilators and two side passages. (Photo by the author, 1996.)*

Similarly, the drive-in corncrib likely derives from the early log transverse-crib, rather than being ancestral to it, as some have proposed (Figures 4.7 and 4.17). The evolution of a full-blown, multipurpose barn from a mere granary makes less sense than the modest alteration of the pre-existing Alpine double-crib.

I also disagree with Kniffen and Glassie on the *place* of origin of the transverse-crib. That hearth, I feel, lay in the Watauga settlements. That area today still contains a noteworthy concentration of Alpine double-crib barns, mostly in frame construction, juxtaposed with large numbers of very similar transverse-cribs, differing only in roof orientation. The Watauga district also has a great many fine, larger, mature transverse-crib barns (Figure 4.18). It is perhaps noteworthy, too, that Watauga is flanked on all sides, and especially to the north in Virginia, by a major concentration of log drive-in corncribs.

Almost certainly, the transverse-crib barn and upland southern mixed farming system arose simultaneously and in the same locale. The place seems to have been the Watauga district and the time the 1790s, roughly a generation after the initial pioneer settlement of the area.[39] After ascending to dominance in Middle Tennessee, the transverse-crib spread throughout the Upland South along paths now familiar to us. In the process it became the single most reliable visible indicator of upper southern presence.

Not all the cultural landscape features that help reveal and explain the Upland South are rural. Some reside in county towns spaced regularly through the region. In the next chapter, I look at the upland southern county seats.

CHAPTER 5

County Seats and Courthouse Squares

Across most of the Upland South, the county seat town remains to this day the single "most important type of urban center."[1] Passing through the region, the traveler must thread through a county seat town every twenty or thirty miles, negotiating a busy courthouse square at the center of almost every one.[2] The square encompasses a rectangular block "surrounded by streets, with the courthouse—often the most ornate building in the county—standing alone in the middle," flanked by lawns and trees (Figure 5.1).[3]

FIGURE 5.1 *A courthouse at Lockhart, Caldwell County, Texas. With a fine structure such as this, a county in the Upland South expressed its civic pride. (Photo by the author, 1987.)*

On all four outer sides of the square, the town's leading businesses occupy rows of contiguous two-story buildings, facing the courthouse. Retail shops, banks, cafes, and artisanal establishments occupy the street level of these rows, along with barber and beauty shops, while doctors, dentists, lawyers, and fraternal lodge halls find places in the upper rooms. A hotel or movie theater occupying both levels is sometimes present (Figure 5.2).

The architects of the courthouses typically took care to design four equal façades for the hall of justice, so as not to favor one commercial row over another.[4] Symmetry reminiscent of the dogtrot house or transverse-crib barn was the order of the day.

FIGURE 5.2 *On the perimeter of the Floresville courthouse square, a "Shelbyville" type, in Wilson County, Texas. (Photo by the author, 1984.)*

Generally absent from the square, banished to more peripheral locations within the town, are churches, schools, and residences. The square is all about the law, justice, and private enterprise—a celebration of American secular democracy and free enterprise.

This image is as old as the Upland South. An English traveler in the middle nineteenth century, journeying on horseback across North Texas, found in the Lamar County seat of Paris "the usual plan, with a Court House in the centre of the square," followed a short distance later by the town of McKinney in Collin County, a twin of Paris.[5] Like the dogtrot house, the county seat's courthouse square became in time an icon of sorts in the Upland South, entering the regional literature in the works of William Faulkner and others.[6]

The County

FIGURE 5.3 *King & Queen Court House, in the Virginia Tidewater district. The courthouse stands alone in a rural area, in contrast to the upland southern tradition of a county seat town. (Photo by the author, 1980.)*

All of this—town, courthouse, and square—rested ultimately upon a decision made early in colonial times that the county (fortunately without counts!) would be the largest administrative unit below the level of the individual colony. The first eight counties in North America were created in 1634 in Virginia, and this governmental institution eventually spread throughout the eastern seaboard, from Massachusetts to Georgia.[7]

But cultural divides developed almost at once. New Englanders soon marginalized the role of the county, emphasizing instead the smaller township, or "town," as the basic administrative building block of government. This allowed an admirable populist democracy resting upon regularly scheduled town meetings held in halls erected for that purpose. In the process, New England went its own way governmentally. It would have no influence upon the upland southern county or its seat.

The Tidewater and Low Country of the southern seaboard also chose their own divergent path. Instead of developing a principal town in which county administration could be joined to other functions, the deeply rural lowland southerners usually erected a freestanding courthouse out in the countryside, at a crossroads. For example, even today the seat of government of King and Queen County, in the Virginia Tidewater, remains at the place aptly called "King and Queen Court House" (Figure 5.3). We cannot, then, look to the southern Low Country for the prototype of the upland southern county seat town. Instead, that distinction belongs to Pennsylvania. The county seat town came to the Upland South from the Delaware Valley primary hearth (Figure 1.5).

The County Seat

In the colonial-derived system, when legislatures established a new county, their action described its boundaries and specified a site for the seat of justice. Later, the decision locating the county seat passed to the residents, opening the way for fierce competition and occasional relocations. In terms of law, all the legislature required was a place to hold court and to confine, punish, or execute convicted criminals. Jails, whipping posts, stocks, "hanging" trees, and rooms for the court and jury sufficed, in the legal sense.

Pennsylvania's primary contribution, probably borrowed from the recently colonized Ulster Plantation in Northern Ireland, was to place the seat of county law in a town, combining the place of justice with commercial activities, services, county administration, and record keeping in one dominant settlement. These activities and functions, in turn, became part of the broader social and community life that characterizes urban places, including churches, schools, clubs, celebrations, parades, public speeches, and festivals.

The Courthouse

Pennsylvanian, too, was the idea that the county courthouse should be a noble structure, one that both paid visual tribute to the evolving American democracy and the power of the law, while also giving expression to civic and county pride. In the Pennsylvania-derived Midland culture, the ornate, professionally designed courthouse played the role once occupied by temples or grand amphitheaters in the classical Mediterranean world—the single most important structure, in which was embodied the achievements and aspirations of the county and its seat town. No church spire or mansion would visually challenge this symbol of secular democracy and capitalistic success.

The Pennsylvanians fabricated all this architectural grandiosity atop the very modest demands placed by law upon the courthouse—that it provide a roof over two rooms, one for the court and another for the jury. The earliest courthouses of the Upland South were typically log buildings, not infrequently of the "dogtrot" plan, with courtroom and jury deliberation quarters occupying the two pens.[8] Only gradually, as administrative offices and record storage found a place under the same roof and the county passed from the frontier era into family-based agrarian prosperity, did the courthouse become a grand structure. In the process, the grounds around the courthouse changed, too. Originally a dusty or muddy expanse where peddlers and farmers sold wares and produce in an open-air market, where county militias drilled and hanged felons occasionally dangled on ropes from sturdy tree limbs, where rowdy gatherings at adjacent taverns sometimes brought the county sheriff, the courthouse grounds in time echoed the beauty and grandeur of the building (Figure 5.4). Well-manicured lawns, towering trees, and flower beds appeared, as did monuments to war dead and famous persons. Fourth of July parades, public speeches by politicians, sedate games of dominoes, and the gerontocracy of the "spit-and-whittle" bench replaced executions, military drills, and snakeoil peddling.

FIGURE 5.4 *The courthouse square at Clinton, Henry County in western Missouri, is a classic example of the "Shelbyville" type that is dominant in the Upland South. (Photo 1975, courtesy Clinton, Missouri, Chamber of Commerce.)*

The Town Square

The upland southern idea that the most suitable place for the courthouse and its surroundings is in the middle of a public square, or "court square" as it is sometimes called in the Upland South, in the center of the county seat, bordered by streets and businesses, also came from Pennsylvania. Northern Irish towns seem to have provided an earlier, European prototype, but in most respects the Pennsylvanian courthouse square broke sharply with Old World tradition, in which a cathedral occupied the central position, surrounded by a paved open-air market, and the perimeter of the square provided a place for the town hall, guild houses, and the residences of the wealthy.

A rigid geometry shapes the upland southern county seat. The courthouse square is, as its name implies, perfectly rectangular, and the entire town typically displays a checkerboard pattern of streets. At the very least, those streets entering the courthouse square run at right angles to it. These county seats did not develop and grow haphazardly. Instead, surveyors platted the towns at the time of settlement, often choosing a point of higher ground for the square, around which the checkerboard of straight avenues meeting at right angles spread (Figure 5.5). This planning and geometry, thoroughly Pennsylvanian, also apparently has roots in the slightly older colonization of Northern Ireland.[9]

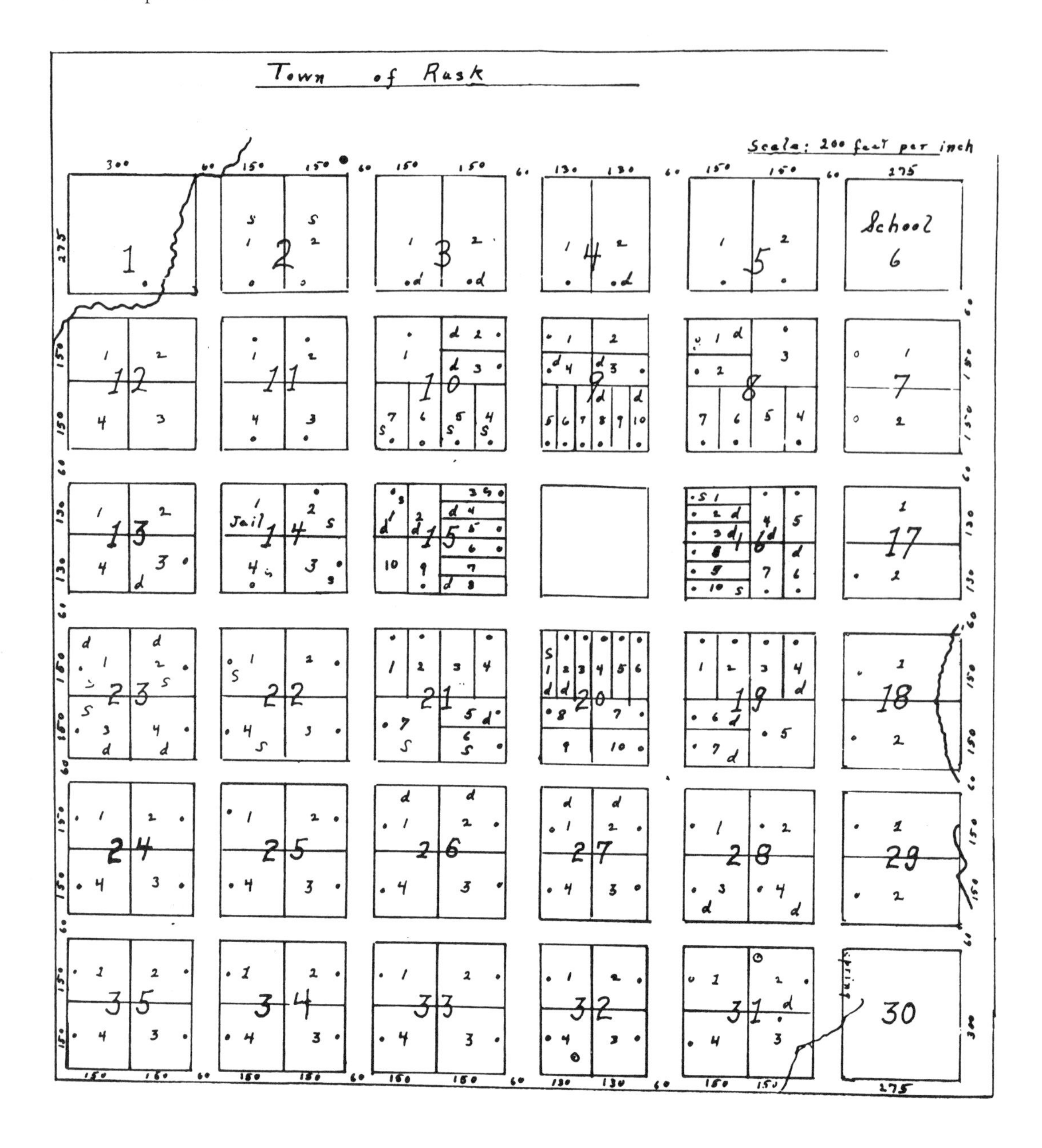

FIGURE 5.5 *A surveyor's original plat of the town of Rusk, seat of Cherokee County, Texas. It provides for a "Shelbyville" type of square. (Copied from the original plat in the General Land Office of the State of Texas, Austin.)*

The Pennsylvania Town

At the same time, the upland southern county seat differs in fundamental ways from the Pennsylvania model, providing the basis of a regional subtype so distinctive as to become another visual index of the Upland South. To illustrate the magnitude of difference between the morphology of the county seats of Pennsylvania and those of its upland southern child, we must first describe what cultural geographer Wilbur Zelinsky aptly called "the Pennsylvania town."[10]

Even a cursory listing of the traits of Pennsylvania towns reveal them to have a very different character from those of the Upland

South. They are compact in structure, a morphology achieved by an affinity for row houses and duplexes, as well as by the virtual elimination of front yards for residences. No clear segregation of dwellings, shops, and offices can be detected. The town square, called a "diamond," is small, with little perimeter space for stores and other businesses, and the county courthouse does not stand in the middle of the square, but instead usually in a corner of the perimeter, further diminishing the space for retail establishments. The diamond remains open and devoid of major structures. Brick prevails as the building material for houses, stores, and public structures, as well as for street paving.[11]

The layout of the Pennsylvania town square itself takes a form rare in the Upland South. Four streets enter the diamond, each at the midpoint of the block, a plan variously referred to as a "Philadelphia," "Lancaster," or "secant" square (Figure 1.17). Philadelphia was platted in 1682 around such a square.[12] The secant square was a British colonial type, and I have seen them in locations as varied as Londonderry, Northern Ireland, and Tanunda, South Australia.

Some such "diamonds" subsequently had a county courthouse erected in the middle, providing a clear prototype for the upland southern custom. In fact, the term "Philadelphia" square is reserved for those which never acquired a central courthouse, while "Lancaster" describes one with a courthouse. The square at Lancaster in southeastern Pennsylvania, was platted in 1729, and a decade later the courthouse was erected there, only to be removed in 1854. Similarly, the town of York, Pennsylvania, was platted in 1741, acquired a central courthouse in 1749, and saw it removed in 1841.[13]

In short, the idea of a public square with a courthouse in the middle first appeared in America in southeastern Pennsylvania, though the plan never became dominant. Then, in the nineteenth century, the courthouses were removed. The parallel to the dogtrot house is exact. Pennsylvania once possessed both the central courthouse square and the dogtrot as minority types, only to discard them. Once again, the Upland South can be interpreted as a repository of archaic Pennsylvanian elements of material culture. As Edward Price said of the courthouse square, Pennsylvania is "the source of a trait it scarcely possesses today."[14] Lancaster, Pennsylvania, in turn, had acquired every single element—county, county seat, Lancaster square, gridiron plan, and central governmental building from Londonderry in Northern Ireland, platted in the early 1600s. There, too, the central building, dating to 1622, was later removed and replaced by a war memorial.[15]

The "Pennsylvania town" never spread significantly any farther than the northern part of the Shenandoah Valley of Virginia (Figure 5.6). Even the Pennsylvanian squares—whether the Philadelphia or Lancaster type—made relatively little headway beyond the boundaries of

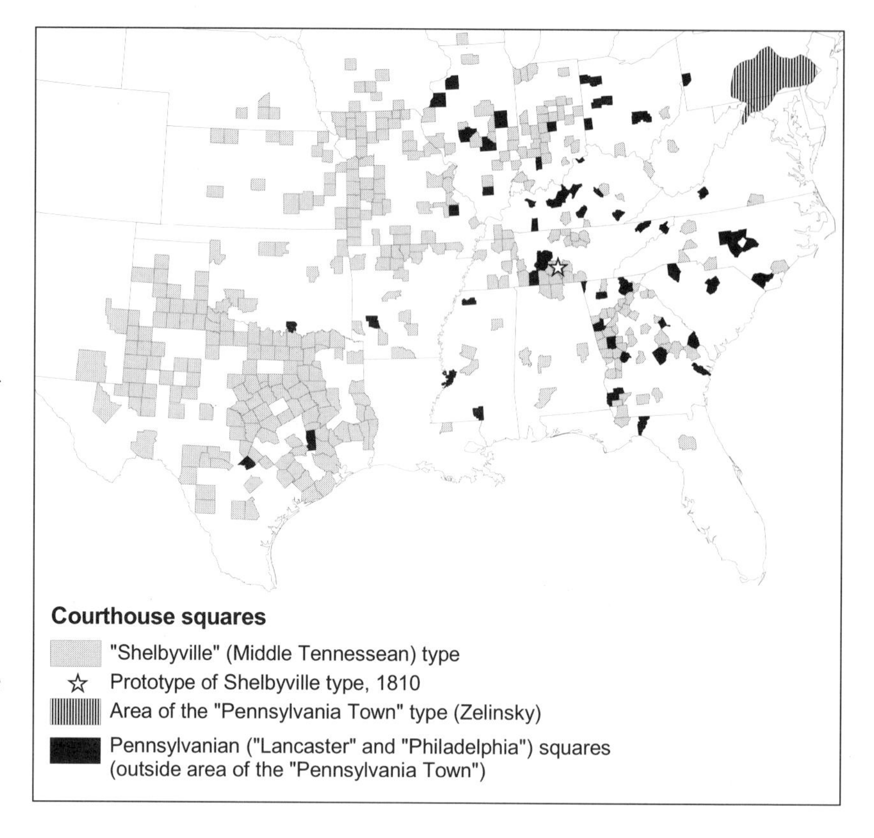

FIGURE 5.6 *Distribution of selected courthouse square types, with an emphasis on the "Shelbyville" type, the prototype of which was reputedly the seat of Bedford County in Middle Tennessee, platted in 1810. Many such squares later underwent modification, usually involving the removal of the courthouse, but this map shows the original configuration. The "Pennsylvania town" essentially failed to penetrate the Upland South. The map is fragmentary, since not all county seats have been inspected. (Sources: Price 1968, 35, 48-49; Price 1996 and 1997; Pillsbury 1968, 183-184; Pillsbury 1978, 119, 121; Zelinsky 1953a, 171; Aikins et al. 1971, Figure 5.6; Zelinsky 1977, 130; Veselka 2000, 25.)*

the Keystone State. A faint trail of such squares tracks across the lower Midwest, and small clusterings and scatterings appear in the Carolinas, Georgia, and Kentucky.[16]

Clearly, the Upland South largely rejected both the Pennsylvania town and its squares. Another kind of town would be created there.

The Shelbyville Square

The county seat towns of the Upland South involve more than merely the preservation of an archaic Pennsylvanian trait. Modification also occurred. It apparently happened in 1810 in Middle Tennessee, the quaternary hearth area, in the final coalescence of upland southern culture. In the town of Shelbyville, seat of Bedford County, a new and distinctly Tennessean layout for the central courthouse square was platted in that year. The "Shelbyville" plan, also called the "central" or "block" type, features streets entering only at the four corners of the square (Figure 1.17).[17] Normally larger than the Philadelphia/Lancaster square, the Shelbyville has more room for businesses along its margins.

Even earlier and farther to the east, upland southerners had rejected other elements of the Pennsylvania town. Their county seats would be loose rather than compact. Businesses and services would be concentrated around the square, with any spillover grouped along some of the eight avenues leading to the square. And, almost invariably, they placed the courthouse in the middle of the square. The result was a distinctive upland southern type of county seat.

Did Shelbyville, Tennessee, really provide the prototype of the square named for it? Not all county seats have been studied, and earlier examples of the Shelbyville plan, lying farther east, could still be found. Richard Pillsbury suggested that two "Shelbyvilles," apparently dating to the 1791–1810 period, occur in the Georgia Piedmont. Indeed, Pillsbury stated that "the concept certainly did not originate in Shelbyville," Tennessee.[18] Still, Edward Price, who studied courthouse squares more extensively than any other scholar, points to Bedford County's square as the earliest example.[19] Given Middle Tennessee's pivotal role in upland southern culture formation and migration, Bedford County makes sense, if not as the place of origin, then at least as the locale where the Shelbyville concept took root in the regional culture and from which it spread. Moreover, Shelbyville and nearby Manchester both contain a church on their respective square perimeter, a feature absent in later examples, suggesting that an early period of experimentation was still underway in 1810.

Spread of the Shelbyville Plan

The appearance of the Shelbyville plan occurred shortly before the migratory explosion out of Middle Tennessee. As a result, Shelbyville squares sprang up all over the extended Upland South (Figure 5.6). In the Piedmont of Georgia, fifty-seven percent of all courthouse squares are "Shelbyvilles," as opposed to only fourteen percent of the Philadelphia/Lancaster type.[20] The splendid square at Forsyth in Monroe County, Georgia, provides an example of the Shelbyville and has been placed on the National Register of Historic Places. Thomaston in adjacent Upson County is another excellent specimen.

Southern Indiana has many such squares, as do western Missouri and the Arkansas Ozarks (Figures 5.4 and 5.6). [21] But more than any other part of the Tennessee Extended, Texas embraced the Shelbyville square (Figure 5.7). The first such town plats appeared in 1836, the very year of Texas independence, and eventually the type dominated a huge block of 114 contiguous counties stretching across the entire state both east-to-west and north-to-south. Spillovers of this belt of Shelbyvilles into adjacent Oklahoma and New Mexico add another four counties to it (Figure 5.6). Many of these squares were later modified, usually by relocation of the courthouse, but enough survive intact in Texas to lend a distinctive upland southern appearance to close to half of the county seats in the state. Sixty-two percent of all squares in Texas are Shelbyvilles or variations of that plan, and some were platted as late as 1900.[22]

FIGURE 5.7 *The courthouse square at Hillsboro in Hill County, Central Texas. The "Shelbyville" plan took the deepest roots in Texas of any state. (Photo 1981 courtesy John L. Bean, a dear and departed friend and colleague.)*

A courthouse square is a place of activity and life. Other special locales in the Upland South have been set aside for the dead, and perhaps it is appropriate, as the study nears its end, to turn to the folk graveyard for additional clues concerning the character, origins, and dispersal of upland southern culture.

Chapter 6

Upland Southern Graveyards

Religion, including mortuary customs, offers a rich field of research for cultural geographers and landscape historians. Not only do religious practices and beliefs vary from one region to another, but also, it has been said, they possess a deep-rooted conservatism which retards change and preserves folk practices. Indeed, religious faith has been acknowledged as a highly conservative aspect of culture and the graveyard as the most conservative facet of religion.[1]

Upland Southern Religion

The Upland South provides both affirmation of these generalizations and refutation. Its religious culture is distinctive, slow to change, and in many ways archaic. But at the same time its practitioners have innovated and modified, profoundly. While belonging to the overarching Protestantism derived from the dissenter chapel-goers of the British Isles, a heritage based in Calvin, Wesley, and Knox, upland southern religious faith possesses a unique combination of inherited, modified, and invented traits.[2]

A great many of its superabundant congregations are independent and unaffiliated with any organized denomination, and in these scattered chapels odd practices such as snake handling, speaking in tongues, ecstasy, and other exuberances abound. Upland southerners invented the camp meeting revival; discarded Calvinistic predestinarianism like the rotten apple it was, embracing frontier free will; and happily stripped away requirements for an educated clergy. This led to unintentional oxymorons such as the "Calvin Free Will Baptist Church" in Calvin, Oklahoma, but these changes produced a purer folk version of Protestantism. When it pleased them, upland southerners founded new denominations to celebrate their religious revisionism, such as Cumberland Presbyterianism, the Campbellites or Disciples of Christ, and the Church of God of Cleveland, Tennessee.

My own third great-grandfather, Alexander Westmoreland (1777–1850), labored as an illiterate Methodist circuit rider in the southern Appalachians. According to an inscription in our yellowed family Bible, Alex died out on the circuit, in the dead of winter, aged, "among strangers and far from home, while proclaiming salvation abroad and calling poor sinners to God."

Their religious revisions completed and satisfied with what they had wrought, upland southerners retreated into a profound conservatism. They even backslid into a mild form of Calvinistic fatalism under the blows of poverty that in time plagued the region. Some have been so naive as to suggest that upland southern Protestantism owes its formative debt to Celtic Britain and Northern Ireland, and particularly to the Scotch-Irish. To accept this simplistic notion is to overlook the distinctiveness, inventiveness, and complexity of this religious culture.[3]

Upland southern religion spread throughout the region at large, borne on the migratory currents of relocation diffusion. It reached northern Louisiana, which "follows the practices common to the Upland South," and spilled into Texas with force.[4] Region and religion remained one.

The Cultural Landscape

As might be expected, upland southern religion left an imprint on the region's cultural landscape. What element might I best choose, what single greater religious artifact, to reveal the Upland South? I could have chosen the traditional camp meeting ground, with its central, open-sided tabernacle or brush arbor and surrounding cabins—an artifact almost certainly borrowed from the Cherokee Indian council meeting grounds.[5]

Or I might have selected the unadorned, steepleless board chapels so typical of the region (Figure 1.2). As its name implies, the chapel, almost invariably painted white, is a modest structure. It consists of a single room, and the lines are clean and simple. The floor plan is rectangular, usually measuring between twenty and thirty feet across the front and between thirty and fifty feet on the sides. Gables face the front and rear of the chapel. Perhaps most often the entrance is a central double door, but frequently two separate single doors spaced some distance apart appear in the front wall. In some sects, the tradition was for men to enter by one of these doors, and women and children by the other. Four clear-glass, square-topped windows admit light and breeze from both sides. Four is definitely the preferred number of windows on the sides, appearing so frequently as to raise the question of some possible religious significance. Another window of the same description is situated behind the pulpit in the back gable wall (Figure 6.1).

In their extreme austerity, lacking any sort of religious symbolism, these chapels express the dissenter Protestant's view of the church structure as merely a place of assembly, not an abode of God or the scene of ritual miracle. The people refer to these chapels as "houses." By this they mean "meeting house" rather than "house of God." The earliest examples of such chapels, built of notched logs, can still be

FIGURE 6.1 *Interior of the Indian Creek Missionary Baptist Church, in the East Cross Timbers of Cooke County, North Texas. The simplicity and austerity do not result from rural poverty, for a natural gas well stands on the church grounds, but instead from the demands of Calvin and Wesley. Movie star Gene Autry, who was born and lies buried nearby, also contributed to its upkeep. His first cousin, Mary Autry Mason, stands at the pulpit. (Photo by the author, 1977.)*

FIGURE 6.2 *A log chapel near Fayetteville in the Arkansas Ozarks. This is the prototypical form of the upland southern "church house." (Photo by the author, 1985.)*

found occasionally in sheltered coves and hollows (Figure 6.2). Unsanctified, as Wesley and Knox demanded, these chapels also serve secular needs as school houses and polling places. The lack of sanctity is also suggested by the names given to these congregations. Only about one in four or five of the chapels has a Biblical or other religious name, such as Mt. Zion, Trinity, Hebron, and New Hope. The remainder bear secular names such as Lonesome Dove, Green Valley, County Line, Shady Grove, Elm Ridge, and Chinn's Chapel.[6]

The Graveyard

Tempting as these chapels and camp meeting grounds are, I have selected another, very different element of the religious cultural landscape, one drawn from what has been called the "Upland South Cemetery."[7] We enter here the realm of the ultra-conservative, seeking explanatory power, for if religion is the most traditional aspect of a culture, then the graveyard represents the most conservative aspect of religion.

An uninitiated visitor to a traditional upland southern graveyard might well be puzzled and astounded. In the nearby white board chapel, as we have seen, the theology of Calvin, Wesley, and Knox discourages visual religious symbolism of almost every kind, even a simple cross and steeple. But beyond the arched gateway of the nearby graveyard, a bewildering variety of symbols, largely pagan in origin, compete for the eye. The symbolism suppressed for centuries in the chapels bursts forth all the more vigorously in the cemetery, making the upland southern folk graveyard a confusing, fascinating, and ultimately revealing place.[8]

An array of distinctive traits characterizes the upland southern cemetery. These include selection of a high point of ground as the site; enclosure by a fence; entry through a formal "lichgate"—literally "corpse gate"—beneath an overarching span that often bears the name of the cemetery; burial in family plots, with feet to the east and husband interred to the right or south of the wife; roofed shelters built over some individual graves or plots; ornamental shrubs, especially roses, crepe myrtles, and cedars or junipers; periodic communal "workings" to maintain the graveyard; removal of grass, or "scraping," to reveal the bare earth; and mounding of graves, upon which rest decorative shells, personal objects belonging to the deceased, Christmas trees, or even food. Sharing the chapels' lack of sanctity, these graveyards need not be next to the church (Figure 6.3).[9] Gravestones are modest in size and epitaphs contain a bare minimum of information. The older markers are often unshaped fieldstones bearing no inscription at all. The aversion to religious symbolism, so evident in the chapels, also claims the tombstones.[10] Another trait of the Upland South cemetery

FIGURE 6.3 *"Graveyard working" at Avinger in Cass County, East Texas, about 1906. The two gravesheds no longer exist. The chapel is Methodist. (Photo courtesy of Janelle Knowles Krumbholz of Dallas, Texas, a descendant of some of the people shown, and used with her kind permission; Anita Walker brought this wonderful picture to my attention.)*

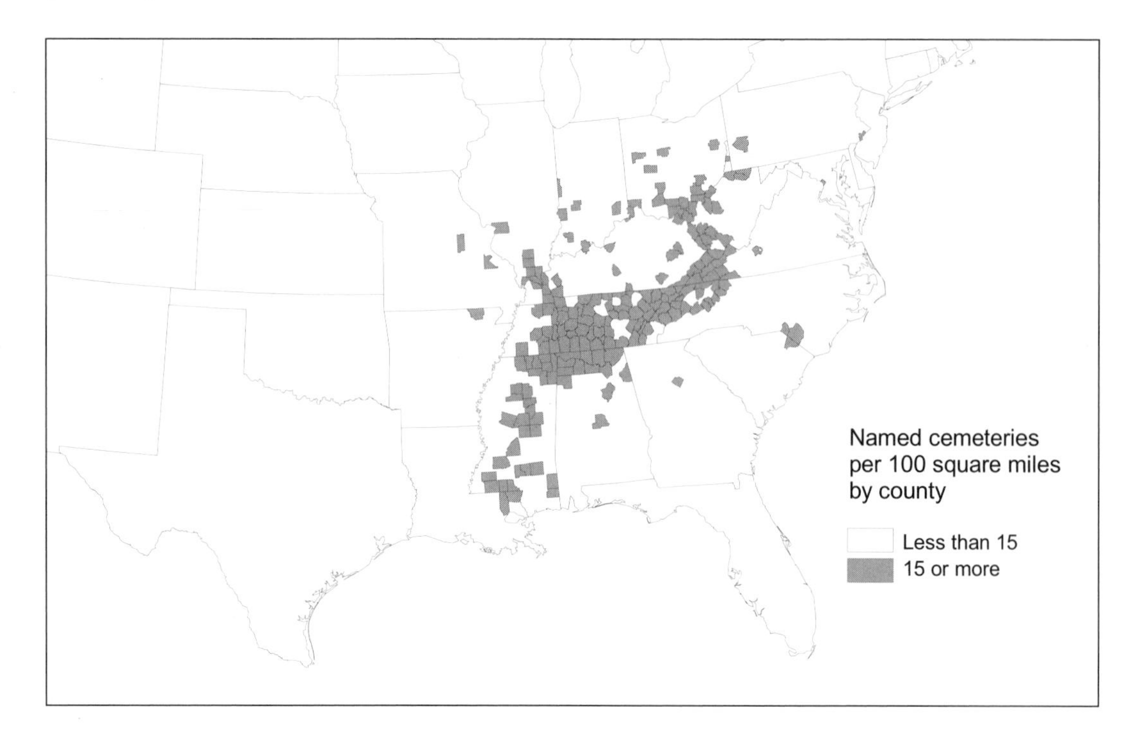

FIGURE 6.4 *Density of named cemeteries in the South. The eastern half of the Upland South stands out as a region of high density, perhaps because small, family graveyards are so common. (Source: simplified and redrawn from Zelinsky 1994, 34.)*

is its remarkable abundance. On a map of graveyards by county, much of the Upland South stands out as the region of greatest density (Figure 6.4). Oddly, this is essentially a characteristic of the cis-Mississippi Upland South.[11]

Gravesheds

Nearly all the traits of the upland southern cemetery seem to have Old World, and particularly European, origin. Many go back to pre-Christian times. These offer us continuity and conservatism but do not provide meaningful insights into the uniqueness of the Upland South.[12] "Scraping" to produce bare-earth expanses offers diagnostic promise, but proves to be not all that universal a practice in the Upland South. It appears to have filtered north from the coastal plain of the Lowland South and may well be an Africanism.[13]

Instead, the most promising artifact is the "graveshed," a small, low, roofed structure with its sides either open or enclosed by pickets, lattices, boards, or wire (Figure 6.5). John Campbell long ago described a cluster of these he happened upon in the southern mountains as "low, latticed houses, painted blue and white."[14] Some of the oldest specimens are built in notched-log construction, with wide, open chinks between.[15] Gravesheds should be distinguished from stone or brick "false crypts," also called "box graves," which have solid walls that conceal the contents.[16] I consider only gravesheds, whose origin in any

FIGURE 6.5 *Gravesheds in the Mills Cemetery, Garland, Dallas County, Texas. Although Anglo-American, these sheds remain true to the original American Indian design in their low profile and picket walls. (Photo by the author, 1977.)*

FIGURE 6.6 *An open-sided graveshed covering multiple burials, in the Ouachita Mountains of Arkansas. The shed is of relatively recent construction. (Photo by the author, 1979.)*

case seems to be independent of the false crypts, though both types cover normal, in-ground burials (Figure 6.6).

The geographical distribution of gravesheds provides another cartographic index to the Upland South (Figure 6.7). As Greg Jeane noted, gravesheds represent a "widespread burial tradition in the Upland

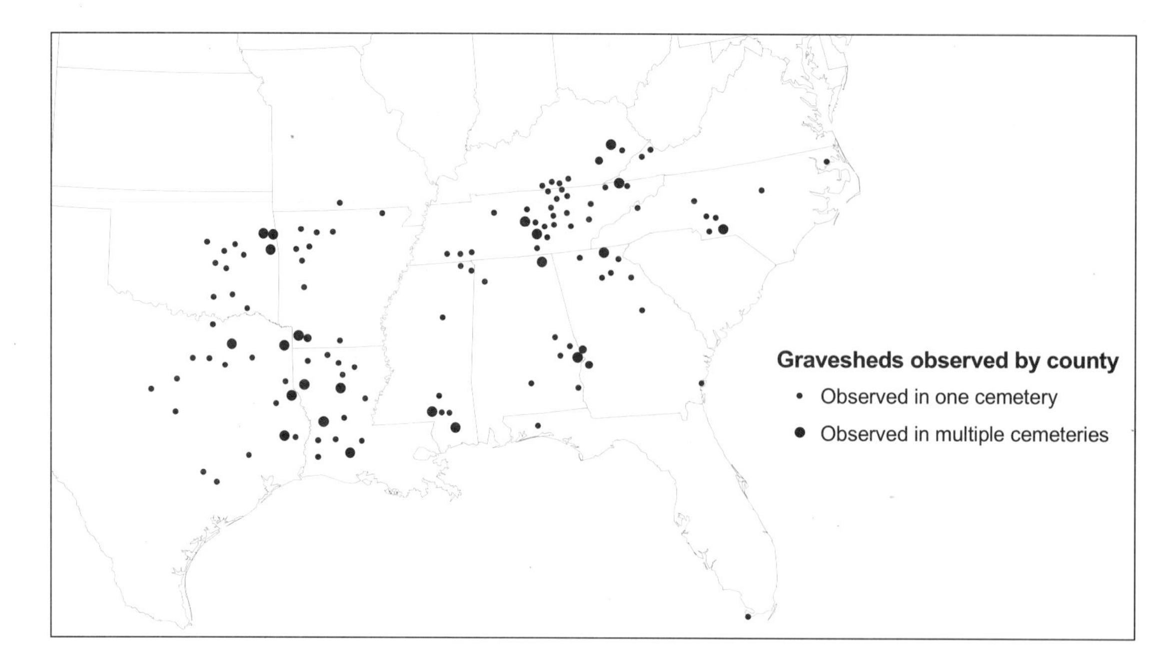

FIGURE 6.7 *Distribution of graveshads observed in the South and bordering areas. The pattern is incomplete, since not all counties have undergone field research. Aside from my own observations, I have relied on numerous works. (Sources: Fielder 1982, 5; Fielder et al. 1977, 31-36; Letcher 1972, 6; Jeane and Purcell 1978, 244–261; Corn 1977; Cobb 1978; Smith 2002; Greiner 2002; Rehder 2002; Morrow 1995; Pitchford 1979, 279; Jeane 1989, 171; Martin 1984, 66; Wright 1956, 148–149; Ball 1977, 30–31; Milbauer 1989b, 178–179; Little 1998, 7–9, 44; Cantrell 1981, 99; Cozzens 1972 whose caption misidentifies the location as near Still*water *instead of Stil*well, *Oklahoma; and for additional sources, see Jordan 1982, 37, and Jordan 1993b, 190.)*

South."[17] Middle Tennessee possesses a concentration of them, as we might expect, but a notable scattering appears to the east, south, and west from there (Figure 6.8).[18] Only the northern margins of the Upland South, including most of Kentucky and Missouri as well as all of Illinois, Indiana, and West Virginia, seem devoid of them. But clearly, the graveshed is distinctively upland southern and must have first appeared there. To determine its origin can teach us something very important about upland southern culture.

While most graveshads stand in the cemeteries of Anglo-Americans, crossing all denominational divides, these shelters are proportionately most common among the surviving remnants of southeastern Indian tribes, most notably in Oklahoma among the Cherokees, Choctaws, Creeks, and Seminoles (Figure 6.9).[19] In the Oklahoma Cherokee counties, John Milbauer found twenty-six graveshads in seven cemeteries he sampled, and they occur most commonly among the so-called "full bloods" and others who retain an ability to speak the Cherokee language.[20] The Cherokees describe the graveshed as "a custom of long ago to protect and comfort the spirit of the deceased."[21] Graveshads also occur among the Kosati, or Coushattas, a subgroup of the Creeks, in Allen Parish, Louisiana.[22] The Melungeons, a mixed-blood "little race" of the southern Appalachians, also built graveshads, especially in Hancock County, Tennessee, but also in "The Breaks," a rugged area in far western Virginia.[23] An elderly Anglo-American man working in the Mt. Zion Cemetery at Doddridge, Miller County, Arkansas, once told me that a graveshed standing nearby represented "an old Indian tradition." Clearly, it seems to me, the graveshed came to upland southern culture from the Indians by way of mixed bloods, a diffusion paralleling

FIGURE 6.8 *A graveshed in the Eagle Springs Baptist Cemetery, Talapoosa County, Alabama. The Anglo-American builder has modified the earlier American Indian design, adding Victorian touches to produce an acculturated version that is widespread among white upland southerners. While this graveshed lies far south of Tennessee, it is in the old Creek Indian homeland, where such structures existed in pre-Christian times. (Photo by the author, 1988.)*

FIGURE 6.9 *Choctaw gravesheds at Homer Chapel Cemetery in Choctaw County, Oklahoma. They retain the original low profile. GraveSheds are most common today in the Indian districts of eastern Oklahoma. (Photo by the author, 1979.)*

that of the camp meeting ground. Tennessee seems to have been the arena of cultural exchange.

Additional support for Indian origin is provided by the absence of gravesheds among non-upland southern whites. You will seek it in vain among the Scotch-Irish of Pennsylvania or Northern Ireland. The English, similarly, have no such tradition. While some gravesheds occur among the Afro-Americans of the southern coastal plain, they are far less likely to build them than either southeastern Indians or upland southern whites. John Milbauer failed to find even one graveshed in a comprehensive field study of cemeteries in the Yazoo Basin of Mississippi—one of the most Africanized parts of the entire South.[24]

An Ancient Custom

John Swanton, a prime authority on southeastern Indians, many years ago found the explanation for gravesheds. Choctaw elders told him that their tribe, in pre-Christian times, placed the deceased atop the ground in the yard of their dwelling and erected a small structure over the dead person, to keep predators away while the body decomposed. The walls consisted of palisades set in the ground, and the roof probably resembled those of brush arbors.[25] Erminie Voegelin implied a similar practice among the Shawnee, who may be the tribe that passed gravesheds on to the Melungeons.[26]

When the body had decomposed, it was removed from the little shed and the bones carefully cleaned by a priest or shaman. The bones then went into a larger charnel house, along with the skeletal remains of many other people. When enough had collected, a sizable burial mound was built as the final resting place.[27]

Conversion to Christianity among the southeastern tribes put an immediate end to this traditional, multi-stage disposition of the dead. Immediate in-ground burial was demanded by the missionaries. The custom of the little houses was allowed to survive, easing the transition into the new religious faith. Only now the graveshed merely protected a burial site from the weather.[28] But the picket walls on many upland southern gravesheds offer a visually startling reminder of their original function.

Once Christianized, the southeastern tribes vigorously intermarried with whites. The graveshed custom passed into the population at large, becoming an aspect of upland southern culture.

This exemplifies a crucially important explanatory mechanism in the upland southern ethnogenesis: indigenousness. With it and the attendant intermarriage came an attachment to region and land perhaps attainable in no other way. It meant "our people have always lived in this place," and that is powerful stuff in the formation of a regional way of life. The Upland South became nobody's child, but instead an independent culture.

Chapter 7

A Region Revealed

In the preceding chapters, five greater artifacts of upland southern material culture have undergone study. They represented five essential attributes of the culture—notched-log carpentry, dwellings, an agro-economic system, county seat towns, and religion. Each artifact was one among many that might have been chosen to depict these five attributes. Forty years of observing the Upland South had led me to believe that the specific examples selected—half-dovetail notching, the "dogtrot" house, the transverse-crib barn, the "Shelbyville" courthouse square, and the graveshed—were each distinctively upland southern and, therefore, had a particular potential to reveal ethnogenesis, adaptation to habitat, cultural diffusion, and the geographic extent of the Upland South. I feel that potential has been demonstrated.

My Concept and Method Reconsidered

The concept of ethnogenesis provides the underpinning of my work. The five artifacts revealed five different processes of ethnogenesis—innovation, selective retention, modification, magnified retention, and indigenousness. These are the processes by which the origin, character, and adaptiveness to a new environment and mix of peoples can be understood, and collectively they reveal the Upland South as a distinctive culture and way of life seated in a distinctive habitat, rather than simply part of the Pennsylvania Extended or the child of several Atlantic seaboard parent cultures.

We need to pay far closer attention to the processes of ethnogenesis, following Leo Gumilev's largely unheeded example.[1] Apparently, diverse peoples often meet along the axis of ecotones, with new cultures created in the process. Certainly, the Upland South developed that way.

(opposite page)
FIGURE 7.1 *The degree of "upper southern-ness" in Texas, 1860. Counties marked with an x were uninhabited at that time, and the dashed line approximates the cultural divide between upland and lowland influences as of that date. (Source: modified from Jordan 1967, 689.)*

The ethnogenetic concept also led me to the notion of four stages of hearth areas (Figure 1.5). The process of ethnogenesis rarely occurs at one particular place and time. Ecotones are transitional zones rather than sharp lines. Moreover, the earliest colonial seaboard hearths all lay in fundamentally different habitats, so that the gatherings of peoples, artifacts, and ideas had not even begun the process of adaptation to the highland habitat.

I also learned that the most crucial hearth area of the Upland South lies in Middle Tennessee. It was there, between about 1790 and 1815, that all of the defining traits came together and achieved general acceptance. It was there that the culture coalesced, precisely as the diverse constituent groups—English, Scotch-Irish, Welsh, Finnish, Swedish, German, and Indian—blended together by intermarriage and lost their older ethnicity, forming one people.

Why did this occur in Middle Tennessee? I really do not know the answer for sure. It probably had something to do with the fact that East Tennessee was not fully subtropical, still displaying many of the climatic tendencies of the continental, cold winters of Pennsylvania and the North, while Middle Tennessee more completely belongs to the subtropics. But I suspect the main reason lay in the reality that the mixing of peoples and all the other ethnogenetic processes just took time, and the frontier had exploded into Middle Tennessee before they ran their course.

Field observation of selected artifacts—my method—is imperfect, of course. Its greatest strength—reliance upon field research and first-hand observation, by me and a host of other scholars—was also its most substantial weakness. Field observations, aside from being highly selective of type of artifact, were spotty, and not every county received equal inspection. More specimens went undetected than entered our field notebooks. But I believe we recorded a representative sample, both numerically and in terms of geographic coverage. Gaps in our collective field research make many maps look motheaten or like Swiss cheese, but the broader aspects of the distribution can be seen.

A Collective Presentation

How might I present the ethnogenetic synthesis of the Upland South in a more general way? A long time ago, when I was young, healthy, and eager to establish myself as a scholar, I published an article depicting the partition of my native state, Texas, between upland and lowland southern cultures.[2] I relied primarily upon agricultural data, throwing in population origins and voting behavior for good measure. At the end of the article I mapped six upland southern traits collectively, as a measure of "upper southern-ness" (Figure 7.1). It seemed to work well enough to demonstrate my point.

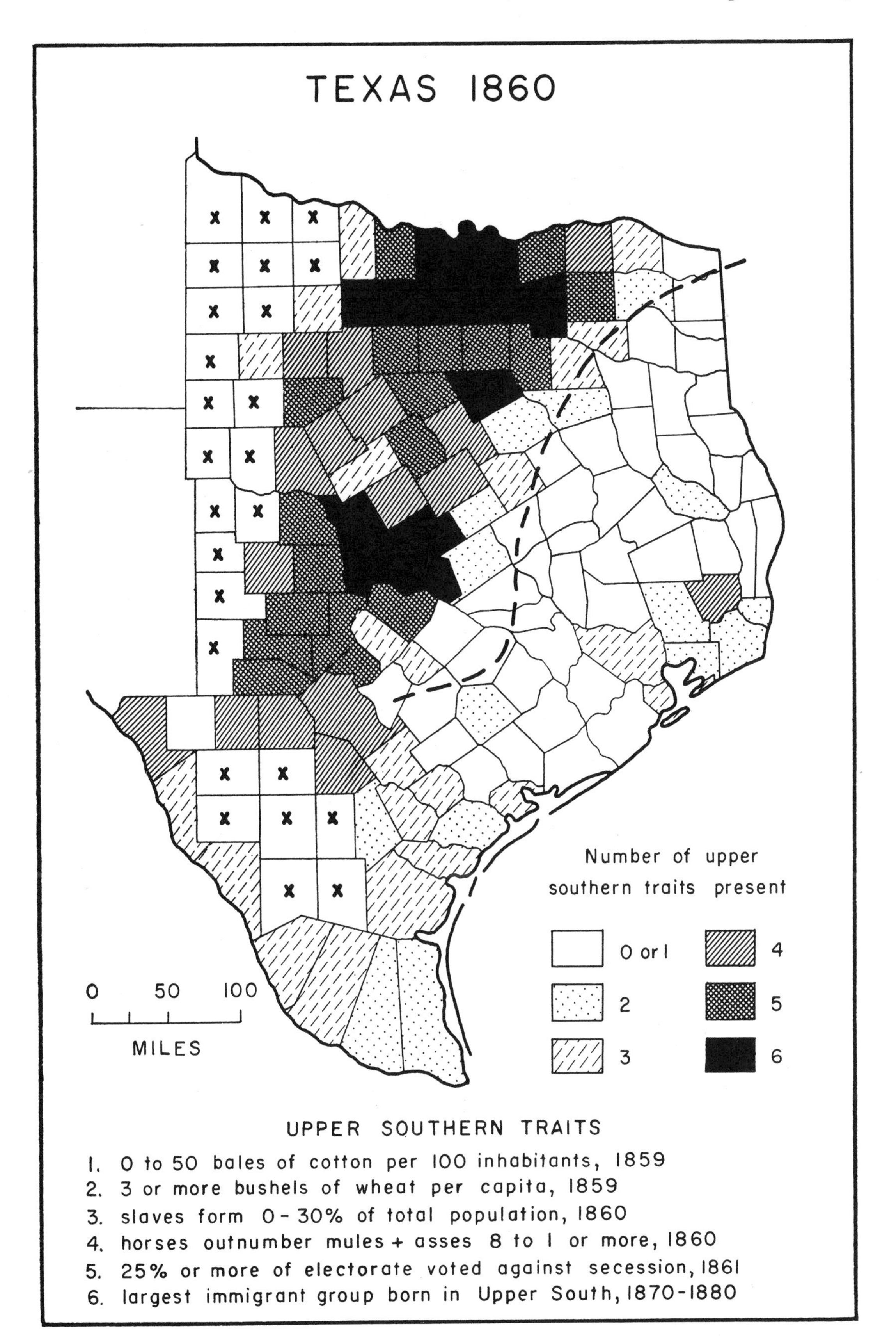
TEXAS 1860
Number of upper southern traits present
0 or 1
2
3
4
5
6
0 50 100
MILES
UPPER SOUTHERN TRAITS
1. 0 to 50 bales of cotton per 100 inhabitants, 1859
2. 3 or more bushels of wheat per capita, 1859
3. slaves form 0-30% of total population, 1860
4. horses outnumber mules + asses 8 to 1 or more, 1860
5. 25% or more of electorate voted against secession, 1861
6. largest immigrant group born in Upper South, 1870-1880

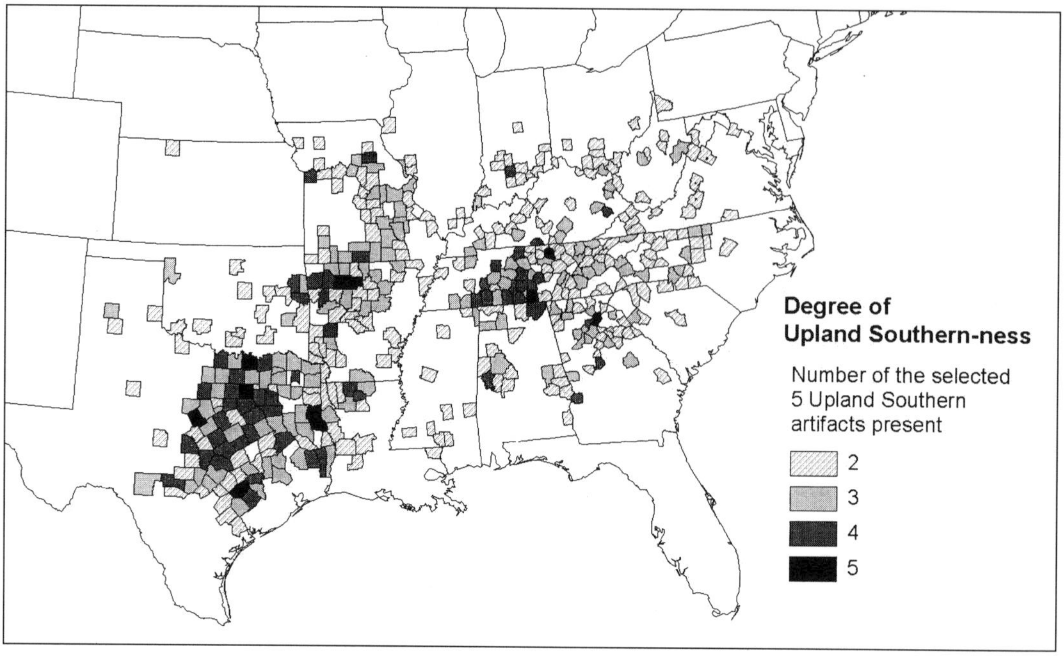

Degree of Upland Southern-ness
Number of the selected 5 Upland Southern artifacts present
2
3
4
5

FIGURE 7.2 *The degree of "upland southern-ness," on the basis of five greater artifacts in the cultural landscape: half-dovetail log notching, the dogtrot log dwelling and open-runway double-crib barn, the transverse-crib barn, the "Shelbyville" courthouse square, and the graveshed. If one or more of these artifacts was observed, the county gets credit for the trait in question. Watch out for moth holes and Swiss cheese. Moreover, the larger size of counties in Texas enhances the likelihood that all or most of the defining traits are present, while Tennessee, Kentucky, and Georgia are penalized for small county size. Mississippi is particularly poorly researched, both by the author and other scholars. The five bases for the map can be seen in figures 2.8, 3.9, 3.12, 4.15, 5.6, and 6.7.*

Now I propose to use the same cartographic method to summarize the preceding five chapters (Figure 7.2). Perhaps the resultant map contains no major surprises, but it possesses the advantage of showing a gradation of Upland Southern influence. No sharp borders exist, as I suggested earlier. Instead, a core/periphery pattern emerges or, better put, a multiple-nuclei configuration that might plagiaristically be called "T for Texas, T for Tennessee," with maybe an "O for the Ozarks" thrown in.

But do not take this cartographic synthesis too literally. Georgia, Kentucky, and Tennessee's smaller counties made it more difficult for multiple traits to occur and be observed. Texas, Arkansas, and Alabama enjoyed the advantage of larger counties. Add to this the problem of spotty field observation and the map becomes merely suggestive. Mississippi, uniquely, has been poorly studied by folklife scholars.

Middle Tennessee, in spite of its small-county problem, clearly reveals its core/hearth function, both because it is densely populated and has benefitted from the meticulous field research of Gerald Smith, John Rehder, and others. Figure 7.2 is a testimony to the role and influence of Middle Tennessee.

Surprises include the weak "signals" received from southern Indiana and Illinois, as well as from most of Kentucky, West Virginia, and Virginia. I suspect these signals are accurate, but I do not know why. Much of the Watauga country, similarly, seems to bear a weak imprint, in spite of the fact that Henry Glassie and many other folklife scholars, including myself, have criss-crossed it numerous times.

Hopefully, a future generation of folklife and landscape students will modify and flesh out this admittedly incomplete map. It should be regarded as an initial proposal rather than a finished product. That might be said of the entire book, though it represents the culmination of much of my personal career.

What I do know for certain is that the Upland South is an indigenous American region of considerable extent and importance. It ranks with the Midwest and West among the most thoroughly American regions.

Notes

Chapter 1

1 McKenzie 1994.

2 Ainsley and Florin 1973; Montell 1975; Price 1960; Wilhelm 1974; Kniffen 1949; Rafferty 1973; Otto and Burns 1981; Eaton 1937; MacClintock 1901; Durand 1956; Miller 1968; Brinkman 1973; Ford 1962; Jordan 1976a; Wilhelm 1993; Cox 1978; Jeane 1978; Jordan and Kaups 1989.

3 Pederson et al. 1986; Carney 1979 and 1996; Hall 1942.

4 Foscue 1945; Carney 1994.

5 Carney 1974 and 1977.

6 Edwards 1935; Owsley 1949; Pudup et al. 1996; Dunn 1988; Mitchell 1991, 284–288, presents an excellent bibliographical overview.

7 Kephart 1916; Thompson 1910; Randolph 1932; Campbell 1921; Upton 1941; Thomas 1930; Lewis 1948; Bryson 1964.

8 Semple 1901; Sauer 1920; Browne 1929; Scofield 1936; Hewes 1940; Hitch 1931; Vance 1932; Rafferty and Hrebec 1973; Wilhelm 1965 and 1978; Zelinsky 1953a; Mitchell 1991; Raitz and Ulack 1984; Mitchell 1977.

9 Kurath 1949; Carver 1986; Jordan 1967; Newton 1974; Shortridge 1995, 2.

10 Zelinsky 1989, 154; Jordan-Bychkov 1998, 6; Newton 1974, 149; Roark 1990, 16; Shortridge 1995, 187, 193; Meyer 2000, 165; Glassie 1968a, 39. See, also, Gordon 1968, 72; Zelinsky 1951; Wilhelm 1972; Meyer 1976b.

11 Meyer 1975 and 1976a; Jordan 1970; Wilhelm 1972; Browne 1929; Clendenen 1973; Johnson 1978; Miller 1968; Otto and Anderson 1982; Rose 1986; Upton 1941; Randolph 1932; Anderson 1937; Barnhart 1935; Crisler 1948; Gerlach 1976; Lang 1954.

12 Baker 1927; Hudson 1994.

13 Pillsbury 1987; Glass 1986; McGreevy 1996.

14 Glassie 1968a, 39.

15 Kniffen 1965; Jordan-Bychkov 1998, 6, 26; Milbauer 1996–97; Kurath 1949; Browning 1915.

16 Shortridge 1995, 2.

17 Braderman 1939; Mitchell 1991, 105–126.

18 Jordan 1993c, 189–199.

19 Meyer 2000, 139.

20 Mitchell 1978, 75; Mitchell 1977; Jordan and Kaups 1989, 9, 10; Newton 1974, 143; Jordan 1993c, 111, 176; Ramsey 1964.

21 Carney 1996, 65.

22 Rehder 1992, 95; Jordan and Kaups 1989, 245; McDonald and McWhiney 1980; Leyburn 1962; Mitchell 1966; Flanders 1986; Berthoff 1986.

23 Semple 1901; Rehder 1992; McWhiney 1988; Gerlach 1986, 13–17; Evans 1965 and 1969; Flanders 1986; Mitchell 1991, 69–86.

24 Gerlach 1984, 47.

25 Allen and Turner 1987.

26 Allen and Turner 1987, 210.

27 Rehder 1992, 103, 105-106; Jordan 1985, 24–25; Glassie 1963 and 1968b.

28 Jordan and Kaups 1989.

29 Mitchell 1991, 87–104; Browning 1915; Glass 1986.

30 Glassie 1968a, 79.

31 Jordan-Bychkov 1998, 12, 24–25.

32 Jordan-Bychkov 1998, 25.

33 Cozzens 1943.

34 Jordan 1993b.

35 Price 1950 and 1951.

36 Hewes 1940 and 1943; Pillsbury 1983; Mitchell 1991, 37–51.

37 Jordan 1993b, 182–190.

38 Otto and Anderson 1982.

39 Meyer 2000, 139.

40 Garrett 1988, 49.

41 Meyer 2000, 140; Anderson 1937.

42 Meyer 2000, 140 (quote); Walz 1952 and 1958; Shoemaker 1955; Gerlach 1976, 178–180; Gerlach 1986, 23, 51–76.

43 Hudson 1988, 400–401; Shoemaker 1955, 130.

44 Barnhart 1935; Meyer 2000, 140.

45 Landis 1923.

46 Meyer 1976a, 158; Meyer 1975, 58; Meyer 2000, 138, 146–151, 288–289; Meyer 1976b, 6–8.

47 Rose 1985, 207; Meyer 2000, 140; Rose 1986, 244; Lang 1954.

48 Browning 1915; Braderman 1939.

49 Treat 1967, 252–259; Grindstaff 1969, 32, 43; Meyer 2000, 140.

50 Jordan 1967, 1969, and 1993a; Kerr 1953; Lathrop 1949; White 1948; Osburn 1963, 305-321; Rather 1904, 189–191.

51 Shortridge 1995, 20, 25, 55, 58, 159, 162.

52 Gumilev 1990; Jordan 1970; Gerlach 1976; Otto and Anderson 1982; Buck 1936; Newton 1974, 143; Sauer 1920; Upton 1941; Rose 1986; Meyer 1975, 57; Letcher 1972.

53 Clevinger 1938 and 1942.

54 Hudson 2000, 74–75; Smith 2002.

55 Gumilev 1990; Hudson 2000, 14.

56 Hudson 2000, 14, 67, 68, 70–71, 74–75; Schroeder 1982, 28; Transeau 1935; Jordan 1973a; Dicken 1935; Rose 1986, 250; Wilhelm 1972, 113; Jordan 1964, 209–211.

57 Bucher 1962; Ensminger 1992.

58 Jordan 1994; Jordan and Kaups 1987 and 1989, 179–191; Hulan 1977.

57 Price 1968; Zelinsky 1977.

58 Jordan-Bychkov 1998.

Chapter 2

1 Wright 1956.

2 Glassie 1964; Withers 1950; Riedl et al. 1976, 20–26; Milbauer 1996–97, 95; Hitch 1931, 310.

3 Jordan 1985, 14-23; Kniffen and Glassie 1966, 53–57.

4 Price 1988; Jordan 1985, 8; Jordan and Kaups 1989, 138–139; Morgan 1990, 59–78; United States Department of Agriculture 1939.

5 Rehder, Morgan, and Medford 1979; Morgan 1990, 79–107.

6 Kniffen 1969; Jordan 1986; Jordan and Kilpinen 1990; Jordan, Kaups, and Lieffort 1986; Jordan, Kaups, and Lieffort 1986–87.

7 Wigginton 1972, 64–75.

8 Jordan 1985, 18–21; Jordan and Kilpinen 1990; Jordan, Kaups, and Lieffort 1986; Jordan, Kaups, and Lieffort 1986–87; Rehder 1992, 103.

9 Kniffen 1969, 3-4; Kniffen and Glassie 1966, 55–56.

10 Gavin 1995, 17.

11 Wigginton 1972, 64–70; Jordan 1973b.

12 Wigginton 1972, 75.

13 Kniffen and Glassie 1966, 56.

14 Kniffen 1969, 3.

15 Wigginton 1972, 66–68.

16 Jordan 1985, 19–20, 47, 57–58.

17 Jordan 1985, 90–94, 120, 131–133.

18 Jordan 1985, 58, 131–134.

19 Glassie 1965a, 50–51 (quote), 59; Gavin 1995, 17; Lyle 1970–74 and 1972; Evans 1966, 76; Wilhelm 1993, 203; Kniffen and Glassie 1966, 56; Mercer 1927, 58.

20 Evans 1969, 78–79.

21 Jordan, Kaups, and Lieffort 1986.

22 Morgan 1990, 35–37; Reding 2002.

23 Gavin 1995, 17; Dickinson 1990, 5; Brumbaugh et al. 1974; Riedl et al. 1976, 27.

24 Roberts 1985, 66; Martin 1984, 19–20, 35–64.

25 Price 1988, 55–56.

26 Jordan 1973b and 1978, 54, 75.

27 Jordan 1978, 54; Milbauer 1996–97.

28 Hutslar 1977 and 1986.

29 Kniffen and Glassie 1966, 61.

30 Alspach 1994, 36, 50–53.

Chapter 3

1 Price 1970, 84.

2 For a fair sampling of this literature, see Zelinsky 1953b; Tebbetts 1978; Jordan 1978; Stuck 1971; Meyer 1975; Milbauer 1996-97; Marshall 1981; Martin 1989; Montell 1976; Morgan 1990; Roberts 1985; Sutherland 1972; Wilson 1975; Wright

1950; Glassie 1963, 1965a, and 1968b; Madden and Jones 1977; Bastian 1977; Ferris 1973; Gavin 1995; Hutslar 1977 and 1986; Scofield 1936.

3 O'Malley and Rehder 1978; Kniffen 1965; Jordan 1985, 23–30.

4 Wilson 1970; Morgan 1990, 20-30. Jordan 1985, 23–25.

5 Riedl et al. 1976, 85–88, 232-234; Jordan 1985, 24, 28; Morgan 1990, 30–32; Dickinson 1990, 7.

6 Smithwick 1983, 169.

7 Morgan 1990, 30–33; Jordan 1985, 24, 28–29.

8 Jordan 1994, references on 47–48; Vlach 1972.

9 Jordan 1994, 37, 44–45; Glassie 1965a, 164–165.

10 Glassie 1965a, 165.

11 Morgan 1990, 30–33; Jordan 1994, 44; Williams 1987, 178, 182.

12 Rehder 2002.

13 Jordan 1994, 44. The builder was John P. Coles, and his birthplace is revealed on a tombstone in the nearby Independence cemetery.

14 Rehder 2002; Dickinson 1990, 7; Gavin 1995.

15 Torma and Wells 1985, 26, 30, 37.

16 Roberts 1985, 128; Vlach 1972.

17 Jordan 1978, 118; Milbauer 1996–97, 94; Martin 1989, 140.

18 Wright 1958; Jordan and Kaups 1987.

19 Hulan 1977; Jordan and Kaups 1989, 179–196.

20 Martin 1984, 38, 109.

21 Flower 1882, 72.

22 Price 1980; Glassie 1969–70; Noble and Cleek 1995, 63–66; Jordan and Kaups 1987, 54–55, 63–64; Jordan and Kaups 1989, 188, 190–191; Roberts 1984, 133–135.

23 Price 1970, 87.

24 Jordan and Kaups 1989, 189.

25 Kniffen 1965, 561; Evans 1969, 80; Jordan and Kaups 1987, 57.

26 Morgan 1990, 31.

27 Hulan 1975; Gavin 1995, 75–81, 104–108.

28 Price 1970; Murray 1982; Gettys 1981; Tebbetts 1978, 47; Wright 1950, 42–49.

29 Evans 1952; Martin 1989, 28–36, 139.

30 Roberts 1985, 126.

31 Pillsbury 1983, 65; Milbauer 1996–97, 93–94.

32 Ferris 1986, 72, 84.

33 Evans 1952; Ferris 1986, 72.

34 Ferris 1986, 74-76.

35 Ferris 1986, 73.

36 Tebbetts 1978, 47.

37 Alexander and Webb 1966, 15.

38 Steinert 1999, 47.

39 Glassie 1965b, 1965-66, and 1969–70.

40 Price 1980; Jordan 1978, 166–173; Milbauer 1996–97, 95–96.

Chapter 4

1 Hudson 1994; Baker 1927; McKenzie 1994.

2 Stewart-Abernathy 1985.

3 Otto and Burns 1981; Arnow 1963; Wilhelm 1966; Newton 1971; Clark 1942; Otto 1983; Owsley 1949.

4 Hart 1977; Otto 1989; Otto and Burns 1981, 173–176.

5 Withers 1950.

6 Hilliard 1969a and 1969b; Durand 1956.

7 Burnett 1946; Wilhelm 1966; Mitchell 1991, 127–149; Jordan 1993c, 189–199.

8 Kniffen 1949; Price 1960; Rafferty 1973; Wilhelm 1974.

9 Jordan and Kaups 1989, 94–134; Arnow 1960a and 1960b; Buck 1936; Johnson 1978.

10 Glassie 1964; Rehder, Morgan, and Medford 1979; Cox 1978; Dunn 1988; Madden and Jones 1977; Noble and Cleek 1995; Schultz 1983a and 1983b; Stewart-Abernathy 1985.

11 Sculle and Price 1993, 23 (quote); Jordan-Bychkov 1998.

12 Sculle and Price 1993.

13 Schimmer and Noble 1984, 26.

14 Hart 1993, 14, 17; Hudson 1994, 102–103.

15 For a list of sources on these barn types, see Jordan-Bychkov 1998, 9, 27–31; see, also, Noble and Cleek 1995, 116–118.

16 Kniffen 1965, 566–567.

17 Glassie 1968a, 92–93; Glassie 1965b, 28–29; Bishir 1990, 300.

18 Wilson 1975, 105; Glassie 1968a, 92.

19 Glassie 1968a, 92; Noble 1977, 73–74.

20 Rehder 1992, 95–99.

21 Sculle and Price 1993, 18.

22 Meyer 1975, 61; Kniffen 1965, 565.

23 Bastian 1977, 131–132.

24 Glassie 1968a, 99.

25 Shortridge 1977, 68–69; Jordan 1978, 169.

26 Riedl et al. 1976, 104–105.

27 Sculle and Price 1993, 23.

28 Moffett and Wodehouse 1993, 8, 12–13, 112; Morgan and Lynch 1984, 95; Glassie 1965–66, 18; Morgan 1997. The Museum of Appalachia, near Norris, Tennessee, displays two such East Tennessee cantilevered forebay barns.

29 Zelinsky 1989, 154.

30 Jordan 1985, 112–113; for additional sources on European central-aisled barns, see Jordan-Bychkov 1998, 20, 29, 31.

31 Noble and Wilhelm 1995, 74.

32 Fitchen 1968.

33 Rehder 1992, 118.

34 Kniffen 1965, 563-564; Glassie 1968a, 88–91.

35 Glassie 1965–66, 16–17.

36 Kniffen 1965, 563–564; Glassie 1968a, 88–91.

37 Glassie 1965b, 29.

38 Jordan 1985, 30-32, 104-105, 108; Martin 1984, 59; Glassie 1965b, 21, 28.

39 Hart 1981, 25. While Hart seems oblivious to the existence of the transverse-crib barn and unwisely tries to find the origins of the Corn Belt in the mountains of southwestern Virginia, he is correct about the importance of that general area to barn evolution. See, also, Hart 1993, 14.

Chapter 5

1 Pillsbury 1978, 115.

2 Price 1968, 29.

3 Price 1968, 29.

4 Price 1968, 29.

5 Barrow 1849, 38–39.

6 Price 1968, 58.

7 Price 1968, 36–37.

8 Jordan 1978, 148–151.

9 Veselka 2000, 1; Pillsbury 1968, 183-185; Price 1968, 35.

10 Zelinsky 1977; Pillsbury 1967.

11 Zelinsky 1977, 130–137.

12 Pillsbury 1967, 116–118; Price 1968, 30, 35, 36, 39; Pillsbury 1978, 119.

13 Price 1968, 40.

14 Price 1968, 41.

15 Price 1968, 36, 41; Evans 1969, 85.

16 Pillsbury 1978, 120, 121; Price 1968, 41–45; Zelinsky 1953a.

17 Price 1968, 44–51; Pillsbury 1978, 119.

18 Pillsbury 1978, 120; Pillsbury 2002.

19 Price 1968, 44.

20 Pillsbury 1978, 118–121.

21 Price 1978, 30, 51.

22 Veselka 2000, 19–20, 23, 32–46, 152; Price 1968, 51; Jordan 1976b, 138.

Chapter 6

1 Jordan 1982, 6.

2 Jordan 1976b.

3 Jordan 1993b, 186–187.

4 Kniffen 1967, 427.

5 Jordan 1993b, 187–188.

6 Jordan 1976a; Roberts 1984, 130–132.

7 Jeane 1969 and 1978.

8 Jordan 1982, 13.

9 Pitchford 1979, 279; Jordan 1980; Jeane 1969 and 1989; Milbauer 1989; Jordan 1982, 13–40; Little 1998, 99–102, 238–241.

10 Jordan 1982, 41–63.

11 Zelinsky 1994, 34.

12 Jordan 1982, 16–34.

13 Jordan 1982, 14–19; Little 1998, 238–241.

14 Campbell 1921, 146–148; Jeane 1989, 170; Ball 1977, 33–41; Corn 1977; Price 1973; Cobb 1978.

15 Fielder et al. 1977, 35–36; Finch 1982, 3.

16 Little 1998, 7, 9; Jordan 1982, 16–17, 20.

17 Jeane 1989, 170.

18 Kniffen 1967, 427; Sexton 1991; Nakagawa 1987.

19 Jordan 1993b, 189; Ball 1977, 30; Cozzens 1972; Jordan 1982, 34; Swanton 1946, plate 88; Zelinsky 1953a, 286.

20 Milbauer 1989a, 178; Hewes 1940, plate 18a.

21 Tyner and Timmons 1970, Volume II, 9

22 Ball 1977, 30.

23 Bible 1975, 107; Ball 1977, 30; Letcher 1972, 6.

24 Milbauer 1989a, 7.

25 Swanton 1931, 183, 185; Swanton 1946, 722, 729, plate 88. See, also, Swanton 1928, 394–397.

26 Voegelin 1944, 272–274, 341, 345.

27 Swanton 1931, 185.

28 Jeane 1989, 171; Ball 1977, 53; Campbell 1921, 148.

Chapter 7

1 Gumilev 1990.

2 Jordan 1967.

Bibliography

Aikins, Douglas B. et al. (eleven others). 1971. *Descriptive and Predictive Modeling of Central Courthouse Square Towns in the South Central United States.* Norman: College of Environmental Design, University of Oklahoma and National Science Foundation (Study GY-9142).

Ainsley, W. Frank, and John W. Florin. 1973. "The North Carolina Piedmont: An Island of Religious Diversity." *West Georgia College Studies in Social Sciences.* 12: 30–34.

Alexander, Drury B., and Todd Webb. 1966. *Texas Homes of the Nineteenth Century.* Austin: University of Texas Press.

Allen, James P., and Eugene J. Turner. 1987. *We the People.* New York: Macmillan.

Alspach, Elizabeth K. 1994. "Reading Ontario's Folk Landscape." M.A. thesis. University of Texas at Austin.

Anderson, Hattie M. 1937. "Missouri, 1804–1828; Peopling a Frontier State." *Missouri Historical Review,* 31: 150–180.

Arnow, Harriette S. 1960a. *Seedtime on the Cumberland.* New York: Macmillan.

Arnow, Harriette S. 1960b. "The Pioneer Farmer and His Crops in the Cumberland Region." *Tennessee Historical Quarterly.* 19: 291–327.

Arnow, Harriette S. 1963. *Flowering of the Cumberland.* New York: Macmillan.

Baker, Oliver E. 1927. "Agricultural Regions of North America: Part IV, The Corn Belt." *Economic Geography.* 3: 447–465.

Ball, Donald B. 1977. "Observations on the Form and Function of Middle Tennessee Gravehouses." *Tennessee Anthropologist.* 2(1): 29–62.

Barnhart, John D. 1935. "Sources of Southern Migration into the Old Northwest." *Mississippi Valley Historical Review.* 22: 49–62

Barrow, John. 1849. *Facts Relating to North-Eastern Texas.* London: Simpkin, Marshall.

Bastian, Robert W. 1977. "Indiana Folk Architecture: A Lower Midwestern Index." *Pioneer America.* 9(2): 115–136.

Berthoff, Rowland. 1986. "Celtic Mist over the South." *Journal of Southern History*. 52: 523–550.

Bible, Jean Patterson. 1975. *Melungeons Yesterday and Today.* Rogersville: East Tennessee Printing.

Bishir, Catherine W. 1990. *North Carolina Architecture*. Chapel Hill: University of North Carolina Press.

Braderman, Eugene M. 1939. "Early Kentucky: Its Virginia Heritage." *South Atlantic Quarterly*. 38: 449–461.

Brinkman, Leonard W. 1973. "Home Manufactures as an Indication of an Emerging Appalachian Subculture." *West Georgia College Studies in the Social Sciences*. 12: 50–58.

Browne, W. A. 1929. "Some Frontier Conditions in the Hilly Portions of the Ozarks." *Journal of Geography*. 28: 181–188.

Browning, Charles H. 1915. "Pennsylvanians in Kentucky." *Pennsylvania Magazine of History and Biography*. 39: 483–484.

Brumbaugh, Thomas B., Martha L. Strayhorn, and Gary B. Gore. 1974. *Architecture of Middle Tennessee: The Historic American Buildings Survey*. Nashville: Vanderbilt University Press.

Bryson, J. Gordon. 1964. *Culture of the Shin Oak Ridge Folk*. Austin: Firm Foundation.

Bucher, Robert C. 1962. "The Continental Log House." *Pennsylvania Folklife*. 12(4): 14–19.

Buck, Solon J. 1936. "Frontier Economy in Southwestern Pennsylvania." *Agricultural History*. 10: 14–24.

Burnett, Edmund C. 1946. "Hog Raising and Hog Driving in the Region of the French Broad River." *Agricultural History*. 20: 86–103.

Campbell, John C. 1921. *The Southern Highlander and His Homeland*. New York: Russell Sage Foundation.

Cantrell, Brent. 1981. "Traditional Grave Structures on the Eastern Highland Rim." *Tennessee Folklore Society Bulletin*. 67(3): 93–103.

Carney, George O. 1974. "Bluegrass Grows All Around: The Spatial Dimensions of a Country Music Style." *Journal of Geography*. 73: 34–55

Carney, George O. 1977. "From Down Home to Uptown: The Diffusion of Country-Music Radio Stations in the United States." *Journal of Geography*. 76: 104–110.

Carney, George O. 1979. "T for Texas, T for Tennessee: The Origins of American Country Music Notables." *Journal of Geography*. 78: 218–225.

Carney, George O. 1994. "Branson: The New Mecca of Country Music." *Journal of Cultural Geography*. 14(2): 17–32.

Carney, George O. 1996. "Western North Carolina: Culture Hearth of Bluegrass Music." *Journal of Cultural Geography*. 16(1): 65–87.

Carver, Craig M. 1986. *American Regional Dialects: A Word Geography*. Ann Arbor: University of Michigan Press.

Clark, Blanche H. 1942. *The Tennessee Yeomen, 1840–1860*. Nashville: Vanderbilt University Press.

Clendenen, Harbert L. 1973. "Settlement Morphology of the Southern Courtois Hills, Missouri, 1820–1860. Ph.D. dissertation. Louisiana State University, Baton Rouge.

Clevinger, Woodrow R. 1938. "The Appalachian Mountaineers in the Upper Cowlitz Basin." *Pacific Northwest Quarterly*. 29: 115–134.

Clevinger, Woodrow R. 1942. "Southern Appalachian Highlanders in Western Washington." *Pacific Northwest Quarterly*. 33: 3–25.

Cobb, James E. 1978. "Supplementary Information on Gravehouses in Tennessee." *Tennessee Anthropological Association Newsletter*. 3(6): 4–7.

Corn, Jack. 1977. "Covered Graves." *Kentucky Folklore Record*. 23(1): 34–37.

Cox, William E. 1978. *Hensley Settlement: A Mountain Community*. N.p. Eastern National Park & Monument Association.

Cozzens, Arthur B. 1943. "Conservation in German Settlements of the Missouri Ozarks." *Geographical Review*. 33: 286–298.

Cozzens, Arthur B. 1972. "A Cherokee Graveyard." *Pioneer America*. 4(1): 8.

Crisler, Robert M. 1948. "Missouri's Little Dixie." *Missouri Historical Review*. 42: 130–139.

Dicken, Samuel N. 1935. "The Kentucky Barrens." *Bulletin of the Geographical Society of Philadelphia*. 33: 42–51.

Dickinson, W. Calvin. 1990. "Log Houses in Overton County, Tennessee." *Tennessee Anthropologist*. 15(1): 1–12.

Dunn, Durwood. 1988. *Cades Cove: The Life and Death of a Southern Appalachian Community, 1818–1937*. Knoxville: University of Tennessee Press.

Durand, Loyal, Jr. 1956. "Mountain Moonshining in East Tennessee." *Geographical Review*. 46: 168–181.

Eaton, Allen H. 1937. *Handicrafts of the Southern Highlands*. New York: Russell Sage Foundation.

Edwards, E. E. 1935. *References on the Mountaineers of the Southern Appalachians.* Washington, D.C.: Government Printing Office.

Elbert, E. Duane, and Keith A. Sculle. 1982. "Log Buildings in Illinois: Their Interpretation and Preservation." *Illinois Preservation Series.* 3: 1–8.

Ensminger, Robert F. 1992. *The Pennsylvania Barn: Its Origins, Evolution, and Distribution in North America.* Baltimore: Johns Hopkins University Press. [A revised and expanded edition was released by John Hopkins University Press in March, 2003.]

Evans, Elliot A. P. 1952. "The East Texas House." *Journal of the Society of Architectural Historians.* 11(4): 1–7.

Evans, E. Estyn. 1965. "The Scotch-Irish in the New World: An Atlantic Heritage." *Royal Society of Antiquaries of Ireland.* 35: 39–49.

Evans, E. Estyn. 1966. "Culture and Land Use in the Old West of North America." *Heidelberger Geographishche Arbeiten.* 15: 72–80.

Evans, E. Estyn. 1969. "The Scotch-Irish: Their Cultural Adaptation and Heritage in the American Old West." In *Essays in Scotch-Irish History*, ed. E. E. R. Green. London: Routledge & Kegan Paul, 69–86.

Ferris, William R., Jr. 1973. "Mississippi Folk Architecture: Two Examples." *Mississippi Folklore Register.* 7:101–114.

Ferris, William R., Jr. 1986. "The Dogtrot: A Mythic Image in Southern Culture." *Southern Quarterly.* 25: 72–85.

Fielder, George F., Jr. (Nick). 1982. "Gravehouses: Mortuary Folk Architecture." *The Courier* (published by Tennessee Historical Commission, Nashville). 21(1): 4–5.

Fielder, George F., Jr. (Nick), Steven R. Ahler, and Benjamin Barrington. 1977. *Historic Sites Reconnaissance of the Oak Ridge Reservation, Oak Ridge, Tennessee.* Oak Ridge: Oak Ridge National Laboratory.

Finch, Richard C. 1982. "Unique Grave Houses in Tennessee." *Standing Stone Press* (newspaper). 4(4) (Spring), 1–3.

Fitchen, John. 1968. *The New World Dutch Barn.* Syracuse: Syracuse University Press.

Flanders, Robert. 1986. "Caledonia: Ozark Legacy of the High Scotch-Irish." *Gateway Heritage.* 6(4): 34–52.

Flower, George. 1882. "History of the English Settlement in Edwards County, Illinois." *Chicago Historical Society Collections.* 1: 30–75

Ford, Thomas R. (ed.). 1962. *Southern Appalachian Region.* Lexington: University of Kentucky Press.

Foscue, Edwin J. 1945. "Gatlinburg: A Mountain Community." *Economic Geography*. 21: 192–205.

Garrett, Wilbur E. (ed.). 1988. *Historical Atlas of the United States*. Washington, D.C.: National Geographic Society.

Gavin, Michael T. 1995. "Nineteenth Century Hewn Log Architecture in Southern Middle Tennessee: An Artifactual Study." M.A. thesis. Middle Tennessee State University, Murfreesboro.

Gerlach, Russel L. 1976. *Immigrants in the Ozarks: A Study in Ethnic Geography*. Columbia: University of Missouri Press.

Gerlach, Russel L. 1984. "The Ozark Scotch-Irish: The Subconscious Persistence of an Ethnic Culture." *Pioneer America Society Transactions*. 7: 47–57.

Gerlach, Russel L. 1986. *Settlement Patterns in Missouri: A Study of Population Origins.* Columbia: University of Missouri Press.

Gettys, Marshall. 1981. "The Dogtrot Log Cabin in Oklahoma." *Outlook in Historic Conservation*. January/February issue: un. p.

Glass, Joseph W. 1986. *The Pennsylvania Culture Region*. Ann Arbor: UMI Research Press.

Glassie, Henry H., III. 1963. "The Appalachian Log Cabin." *Mountain Life and Work*. 39(4): 5–14.

Glassie, Henry H., III. 1964. "The Smaller Outbuildings of the Southern Mountains." *Mountain Life and Work*. 40 (1): 21–25.

Glassie, Henry H., III. 1965a. "Southern Mountain Houses: A Study in American Folk Culture." M.A. thesis. State University of New York, Oneonta.

Glassie, Henry H., III. 1965b. "The Old Barns of Appalachia." *Mountain Life and Work*. 41(2): 21–30 & cover illustration.

Glassie, Henry H., III. 1965–1966. "The Pennsylvania Barn in the South." *Pennsylvania Folklife.* 15(2): 8–19.

Glassie, Henry H., III. 1968a. *Pattern in the Material Folk Culture of the Eastern United States.* Philadelphia: University of Pennsylvania Press.

Glassie, Henry. 1968b. "The Types of the Southern Mountain Cabin." In Jan H. Brunvand (ed.). *The Study of American Folklore: An Introduction*. New York: W. W. Norton, 338–370.

Glassie, Henry H., III. 1969–1970. "The Double-Crib Barn in South-Central Pennsylvania." *Pioneer America*. 1(1): 9–16; 1(2): 40–45; 2(1): 47–52; 2(2): 23–34.

Gordon, Michael H. 1968. "The Upland Southern-Lowland Southern Culture Areas: A Field Study of Building Characteristics in Southern Virginia." M.S. thesis. Rutgers University, New Brunswick.

Greiner, Alyson L. 2002. Letter to Terry G. Jordan-Bychkov. Stillwater, OK, March 27th.

Grindstaff, Carl F. 1969. "Migration and Mississippi." M.A. thesis, University of Massachusett, Amherst.

Gumilev, Leo. 1990. *Ethnogenesis and the Biosphere*. Moscow, Russia: Progress Publishers.

Hall, J. S. 1942. *The Phonetics of Great Smoky Mountain Speech*. New York: King's Crown Press.

Hart, John F. 1977. "Land Rotation in Appalachia." *Geographical Review*. 67: 148–166.

Hart, John F. 1981. "Barns of Southwestern Virginia." *Cela Forum*. 1(1): 25–26.

Hart, John F. 1993. "Types of Barns in the Eastern United States." *Focus*. 43(1): 8–17.

Hewes, Leslie. 1940. "The Geography of the Cherokee Country of Oklahoma." Ph.D. dissertation. University of California, Berkeley.

Hewes, Leslie. 1943. "Cultural Fault Line in the Cherokee Country." *Economic Geography*. 19: 136–142.

Hilliard, Sam B. 1969a. "Hog Meat and Cornpone: Food Habits in the Ante-Bellum South." *Proceedings of the American Philosophical Society*. 113(1): 1–13.

Hilliard, Sam B. 1969b. "Pork in the Ante-Bellum South: The Geography of Self-Sufficiency." *Annals of the Association of American Geographers*. 59: 461–480.

Hitch, Margaret A. 1931. "Life in a Blue Ridge Hollow." *Journal of Geography*. 30: 309–322.

Hudson, John C. 1988. "North American Origins of Middlewestern Frontier Populations." *Annals of the Association of American Geographers*. 78: 395–413.

Hudson, John C. 1994. *Making the Corn Belt*. Bloomington: Indiana University Press.

Hudson, John C. (ed.). 2000. *Goode's World Atlas*, 20th ed. Chicago: Rand McNally.

Hulan, Richard H. 1975. "Middle Tennessee and the Dogtrot House." *Pioneer America*. 7(2): 37–46.

Hulan, Richard H. 1977. "The Dogtrot House and its Pennsylvania Associations." *Pennsylvania Folklife*. 26(4): 25–32.

Hutslar, Donald A. 1977. *The Log Architecture of Ohio*. Columbus: Ohio Historical Society.

Hutslar, Donald A. 1986. *The Architecture of Migration: Log Construction in the Ohio Country, 1750–1850*. Athens: Ohio University Press.

Jeane, D. Gregory. 1969. "The Traditional Upland South Cemetery." *Landscape*. 18(2): 39–42.

Jeane, D. Gregory. 1978. "The Upland South Cemetery." *Journal of Popular Culture*. 11: 895–903.

Jeane, D. Gregory. 1989. "Folk Art in Rural Southern Cemeteries." *Southern Folklore*. 46: 159–174.

Jeane, D. Gregory, and Douglas C. Purcell (eds.). 1978. *The Architectural Legacy of the Lower Chattahoochee Valley in Alabama and Georgia*. University: University of Alabama Press.

Johnson, Oliver. 1978. *A Home in the Woods: Pioneer Life in Indiana. Oliver Johnson's Reminiscences of Early Marion County* (ed. Howard Johnson). Bloomington: Indiana University Press.

Johnston, David E. 1906. *History of Middle New River Settlements and Contiguous Territory*. Huntington, WV: Standard Printing & Publishing Co.

Jordan, Terry G. 1964. "Between the Forest and the Prairie." *Agricultural History*. 38: 205–216.

Jordan, Terry G. 1967. "The Imprint of the Upper and Lower South on Mid-Nineteenth-Century Texas." *Annals of the Association of American Geographers*. 57: 667–690.

Jordan, Terry G. 1969. "Population Origins in Texas, 1850." *Geographical Review*. 59: 83–103.

Jordan, Terry G. 1970. "The Texan Appalachia." *Annals of the Association of American Geographers*. 60: 409–427.

Jordan, Terry G. 1973a. "Pioneer Evaluation of Vegetation in Texas." *Southwestern Historical Quarterly*. 76: 233–254.

Jordan, Terry G. 1973b. "Log Construction in the East Cross Timbers of Texas." *Pioneer America Society Proceedings*. 2: 107–124.

Jordan, Terry G. 1976a. "The Traditional Southern Rural Chapel in Texas." *Ecumene* (published by East Texas State University). 8: 6–17.

Jordan, Terry G. 1976b. "Forest Folk, Prairie Folk: Rural Religious Cultures in North Texas." *Southwestern Historical Quarterly*. 80: 135–162.

Jordan, Terry G. 1978. *Texas Log Buildings: A Folk Architecture.* Austin: University of Texas Press.

Jordan, Terry G. 1980. "The Roses So Red and the Lilies So Fair: Southern Folk Cemeteries in Texas." *Southwestern Historical Quarterly.* 83: 227–258.

Jordan, Terry G. 1982. *Texas Graveyards: A Cultural Legacy.* Austin: University of Texas Press.

Jordan, Terry G. 1985. *American Log Buildings.* Chapel Hill: University of North Carolina Press.

Jordan, Terry G. 1993a. "The Anglo-Texan Homeland." *Journal of Cultural Geography.* 13 (2): 75–86.

Jordan, Terry G. 1993b. "The Anglo-American Mestizos and Traditional Southern Regionalism." *Geoscience and Man.* 32: 175–195.

Jordan, Terry G. 1993c. *North American Cattle Ranching Frontiers: Origins, Diffusion, and Differentiation.* Albuquerque: University of New Mexico Press.

Jordan, Terry G. 1994. "The Saddlebag House Type and Pennsylvania Extended." *Pennsylvania Folklife.* 44(1): 36–48.

Jordan, Terry G., and Matti Kaups. 1987. "Folk Architecture in Cultural and Ecological Context." *Geographical Review.* 77: 52–75.

Jordan, Terry G., and Matti Kaups. 1989. *The American Backwoods Frontier: An Ethnic and Ecological Interpretation.* Baltimore: Johns Hopkins University Press.

Jordan, Terry G., Matti Kaups, and Richard M. Lieffort. 1986. "New Evidence on the European Origin of Pennsylvanian V Notching." *Pennsylvania Folklife.* 36(1): 20–31.

Jordan, Terry G., Matti Kaups, and Richard M. Lieffort. 1986–87. "Diamond Notching in America and Europe." *Pennsylvania Folklife.* 36(2): 70–78.

Jordan, Terry G., and Jon T. Kilpinen. 1990. "Square Notching in the Log Carpentry Tradition of Pennsylvania Extended." *Pennsylvania Folklife.* 40(1): 2–18.

Jordan-Bychkov, Terry G. 1998. "Transverse-Crib Barns, the Upland South, and Pennsylvania Extended." *Material Culture.* 30(2): 1–31.

Kephart, Horace. 1916. *Our Southern Highlanders.* New York: Outing Publishing Co.

Kerr, Homer L. 1953. "Migration into Texas, 1865–1880." Ph.D. dissertation. University of Texas, Austin.

Kniffen, Fred B. 1949. "The Deer-Hunt Complex in Louisiana." *Journal of American Folklore*. 62: 187–188.

Kniffen, Fred B. 1965. "Folk Housing: Key to Diffusion." *Annals of the Association of American Geographers*. 55: 549–577.

Kniffen, Fred B. 1967. "Necrogeography in the United States." *Geographical Review*. 57: 426–427.

Kniffen, Fred B. 1969. "On Corner Timbering." *Pioneer America*. 1(1): 1–8.

Kniffen, Fred B., and Henry Glassie. 1966. "Building in Wood in the Eastern United States: A Time-Place Perspective." *Geographical Review*. 56: 40–66.

Kurath, Hans. 1949. *A Word Geography of the Eastern United States*. Ann Arbor: University of Michigan Press.

Landis, Edward B. 1923. "The Influence of Tennesseans in the Formation of Illinois." *Transactions, Illinois State Historical Society*. 30: 133–153.

Lang, Elfrieda. 1954. "Southern Migration to Northern Indiana before 1850." *Indiana Magazine of History*. 50: 349–356.

Lathrop, Barnes F. 1949. *Migration into East Texas 1835–1860*. Austin: Texas State Historical Association.

Letcher, Peter M. 1972. "The Breaks, Virginia." *Pioneer America*. 4(2): 1–7.

Lewis, Oscar. 1948. *On the Edge of the Black Waxy: A Cultural Survey of Bell County, Texas*. St. Louis: Washington University.

Leyburn, James G. 1962. *The Scotch-Irish: A Social History*. Chapel Hill: University of North Carolina Press.

Little, M. Ruth. 1998. *Sticks and Stones: Three Centuries of North Carolina Gravemarkers*. Chapel Hill: University of North Carolina Press.

Lyle, Royster, Jr. 1970–1974. "Log Buildings in Rockbridge County." *Proceedings of the Rockbridge Historical Society*. 8: 3–12.

Lyle, Royster, Jr. 1972. "Log Buildings in the Valley of Virginia." *Roanoke Valley Historical Society Journal*. 8(1): 24–31.

MacClintock, S. S. 1901. "Kentucky Mountains and Their Feuds." *American Journal of Sociology*. 7: 1–28, 171–187.

McDonald, Forrest, and Grady McWhiney. 1980. "The Celtic South." *History Today*. 30: 11–15.

McGreevy, Patrick. 1996. "Allegheny Identity: Regional Distinction in Western Pennsylvania." *Abstracts, 92nd Annual Meeting, Association of American Geographers, Charlotte, North Carolina*. Washington, D.C.: Association of American Geographers. 196.

McKenzie, Robert T. 1994. *One South or Many? Plantation Belt and Upcountry in Civil War-Era Tennessee*. New York: Cambridge University Press.

McWhiney, Grady. 1988. *Cracker Culture: Celtic Ways in the Old South*. University: University of Alabama Press.

Madden, Robert R., and T. Russell Jones. 1977. *Mountain Home: The Walker Family Farmstead, Great Smoky Mountains National Park*. Washington, D.C.: U.S. Department of the Interior, National Park Service.

Marshall, Howard W. 1981. *Folk Architecture in Little Dixie: A Regional Culture in Missouri*. Columbia: University of Missouri Press.

Martin, Charles E. 1984. *Hollybush: Folk Building and Social Change in an Appalachian Community*. Knoxville: University of Tennessee Press.

Martin, F. Lestar. 1989. *Folk and Styled Architecture in North Louisiana: Volume I, The Hill Parishes*. Lafayette: Center for Louisiana Studies, University of Southwestern Louisiana.

Mercer, Henry C. 1927. "The Origin of Log Houses in the United States." *Old-Time*

New England. 18 (2): 51–63.

Meyer, Douglas K. 1975. "Diffusion of Upland South Folk Housing to the Shawnee Hills of Southern Illinois." *Pioneer America*. 7(2): 56–66.

Meyer, Douglas K. 1976a. "Southern Illinois Migration Fields: The Shawnee Hills in 1850." *Professional Geographer*. 28: 151–160.

Meyer, Douglas K. 1976b. "Illinois Culture Regions at Mid-Nineteenth Century." *Bulletin of the Illinois Geographical Society*. 18(2): 3–13.

Meyer, Douglas K. 2000. *Making the Heartland Quilt: A Geographical History of Settlement and Migration in Early Nineteenth-Century Illinois*. Carbondale: Southern Illinois University Press.

Milbauer, John A. 1989a. "Southern Folk Traits in the Cemeteries of Northeastern Oklahoma." *Southern Folklore*. 46: 175–185.

Milbauer, John A. 1989b. "Cemeteries of Mississippi's Yazoo Basin." *Mid-South Geographer*. 5: 1–19.

Milbauer, John A. 1996–97. "Pennsylvania Extended in the Cherokee Country: A Study of Log Architecture." *Pennsylvania Folklife*. 46: 92–101.

Miller, E. Joan Wilson. 1968. "The Ozark Culture Region as Revealed by Traditional Materials." *Annals of the Association of American Geographers*. 58: 51–77.

Mitchell, Robert D. 1966. "The Presbyterian Church as an Indicator of Westward Expansion in 18th Century America." *Professional Geographer*. 18: 293–299.

Mitchell, Robert D. 1977. *Commercialism and Frontier: Perspectives on the Early Shenandoah Valley.* Charlottesville: University Press of Virginia.

Mitchell, Robert D. 1978. "The Formation of Early American Cultural Regions: An Interpretation." In *European Settlement and Development in North America: Essays on Geographical Change in Honour and Memory of Andrew Hill Clark*, ed. James R. Gibson. Toronto: University of Toronto Press, 66–90.

Mitchell, Robert D. (ed.) 1991. *Appalachian Frontiers: Settlement, Society and Development in the Preindustrial Era.* Lexington: University Press of Kentucky.

Moffett, Marian, and Lawrence Wodehouse. 1993. *East Tennessee Cantilever Barns.* Knoxville: University of Tennessee Press.

Montell, W. Lynwood. 1975. *Ghosts along the Cumberland: Deathlore in the Kentucky Foothills.* Knoxville: University of Tennessee Press.

Montell, W. Lynwood, and Michael L. Morse. 1976. *Kentucky Folk Architecture.* Lexington: University Press of Kentucky.

Morgan, John. 1990. *The Log House in East Tennessee.* Knoxville: University of Tennessee Press.

Morgan, John. 1997. "The Cantilever Barn in Southwest Virginia." In Michael J. Puglisi (ed.). *Diversity and Accommodation: Essays on the Cultural Composition of the Virginia Frontier.* Knoxville: University of Tennessee Press, 275–294.

Morgan, John, and Ashby Lynch, Jr. 1984. "The Log Barns of Blount County, Tennessee." *Tennessee Anthropologist*. 9(2): 85–103.

Morrow, Lynn. 1995. Letter to Terry G. Jordan-Bychkov. Jefferson, MO, September 22nd.

Murray, Ruth C. 1982. "The Absalom Autrey Log House, Lincoln Parish, Louisiana." *Pioneer America*. 14: 137–140.

Nakagawa, Tadashi. 1987. "The Cemetery as a Cultural Manifestation: Louisiana Necrogeography." PhD. Dissertation. Louisiana State University, Baton Rouge.

Newton, Milton B. 1971. "The Annual Round in the Upland South." *Pioneer America*. 3(2): 63–73.

Newton, Milton B. 1974. "Cultural Preadaptation and the Upland South." *Geoscience and Man*. 5: 143–154.

Noble, Allen G. 1977. "Barns as Elements in the Settlement Landscape of Ohio." *Pioneer America*. 9(1): 63–79.

Noble, Allen G., and Richard K. Cleek. 1995. *The Old Barn Book: A Field Guide to North American Barns and Other Farm Structures.* New Brunswick: Rutgers University Press.

Noble, Allen G., and Hubert G. H. Wilhelm. 1995. *The Barns of the Midwest.* Bloomington: Indiana University Press.

O'Malley, James R., and John B. Rehder. 1978. "The Two-Story Log House in the Upland South." *Journal of Popular Culture*. 11(4): 904–915.

Osburn, Mary M. (ed.). 1963. "The Atascosita Census of 1826." *Texana*. 1: 299–321.

Otto, John S. 1983. "Southern Plain Folk Agriculture: A Reconsideration." *Plantation Society in the Americas*. 2(1): 29–36.

Otto, John S. 1989. "Forest Fallowing in the Southern Appalachian Mountains: A Problem in Comparative Agricultural History." *Proceedings of the American Philosophical Society.* 133: 51–63.

Otto, John S., and Nain E. Anderson. 1982. "The Diffusion of Upland South Folk Culture, 1790–1840." *Southeastern Geographer*. 22: 89–98.

Otto, John S., and Augustus M. Burns, III. 1981. "Traditional Agricultural Practices in the Arkansas Highlands." *Journal of American Folklore*. 94: 166–187.

Owsley, Frank. 1949. *Plain Folk of the Old South*. Baton Rouge: Louisiana State University Press.

Paddock, B. B. (ed.). 1906. *A Twentieth Century History and Biographical Record of North and West Texas.* Chicago and New York: Lewis Publishing Co.

Pederson, Lee, Susan L. McDaniel, Guy Bailey, Marvin Bassett, and Carol M. Adams (eds.). 1986. *Linguistic Atlas of the Gulf States*. 2 vols. Athens: University of Georgia Press.

Pillsbury, Richard R. 1967. "The Market or Public Square in Pennsylvania, 1682–1820." *Proceedings of the Pennsylvania Academy of Science*. 41: 116–118.

Pillsbury, Richard R. 1968. "The Urban Street Patterns of Pennsylvania Before 1815: A Study in Cultural Geography." Ph.D. dissertation. Pennsylvania State University, State College.

Pillsbury, Richard R. 1978. "The Morphology of the Piedmont Georgia County Seat Before 1860." *Southeastern Geographer*. 18: 115–124.

Pillsbury, Richard R. 1983. "The Europeanization of the Cherokee Settlement Landscape Prior to Removal: A Georgia Case Study." *Geoscience and Man*. 23: 59–69.

Pillsbury, Richard R. 1987. "The Pennsylvania Culture Area: A Reappraisal." *North American Culture*. 3(2): 37–54.

Pillsbury, Richard R. 2002. Letter to Terry G. Jordan-Bychkov. Charleston, SC, April 5th.

Pitchford, Anita. 1979. "The Material Culture of the Traditional East Texas Graveyard." *Southern Folklore Quarterly*. 43: 277–290.

Price, Beulah M. D. 1970. "The Dog-Trot Log Cabin: A Development in American Folk Architecture." *Mississippi Folklore Register*. 4(3): 84–89.

Price, Beulah M. D. 1973. "The Custom of Providing Shelter for Graves." *Mississippi Folklore Register*. 3(1): 8–10.

Price, Edward T. 1950. "Mixed-Blood Populations of Eastern United States as to Origins, Locations, and Persistence." Ph.D. dissertation. University of California, Berkeley.

Price, Edward T. 1951. "The Melungeons: A Mixed-Blood Strain of the Southern Appalachians." *Geographical Review*. 41: 256–271.

Price, Edward T. 1960. "Root Digging in the Appalachians: The Geography of Botanical Drugs." *Geographical Review*. 50: 1–20.

Price, Edward T. 1968. "The Central Courthouse Square in the American County Seat." *Geographical Review*. 58: 29–60.

Price, Edward T. 1996. Letter to Terry G. Jordan-Bychkov. Eugene, OR, September 5th.

Price, Edward T. 1997. Letter to Terry G. Jordan-Bychkov. Eugene, OR, January 16th.

Price, H. Wayne. 1980. "The Double-Crib Log Barns of Calhoun County." *Journal of the Illinois State Historical Society*. 73(2): 140–160.

Price, H. Wayne. 1988. "The Persistence of a Tradition: Log Architecture of Jersey County, Illinois." *Pioneer America Society Transactions*. 11:54–62.

Pudup, Mary Beth, Dwight B. Billings, and Altina L. Waller (eds.). 1995. *Appalachia in the Making: The Mountain South in the Nineteenth Century*. Chapel Hill: University of North Carolina Press.

Rafferty, Milton D. 1973. "The Black Walnut Industry: The Modernization of a Pioneer Custom." *Pioneer America*. 5(1): 23–32.

Rafferty, Milton D. 2001. *The Ozarks: Land and Life*. 2nd ed. Fayetteville: University of Arkansas Press.

Rafferty, Milton D., and Dennis J. Hrebec. 1973. "Logan Creek: A Missouri Ozark Valley Revisited." *Journal of Geography*. 72(7): 7–17.

Raitz, Karl B., and Richard Ulack. 1984. *Appalachia: A Regional Geography*. Boulder: Westview Press.

Ramsey, Robert W. 1964. *Carolina Cradle: Settlement of the Northwest Carolina Cradle*. Chapel Hill: University of North Carolina Press.

Randolph, Vance. 1932. *Ozark Mountain Folks*. New York: Vanguard.

Rather, Ethel Z. 1904. "DeWitt's Colony." *Texas State Historical Association Quarterly*. 8: 95–192.

Reding, William M. 2002. "Assessment of Spatial and Temporal Patterns of Log Structures in East Tennessee." M. S. thesis. University of Tennessee, Knoxville.

Rehder, John B. 1992. "The Scotch-Irish and English in Appalachia." In Allen G. Noble (ed.). *To Build in a New Land*. Baltimore: Johns Hopkins University Press, 95–118.

Rehder, John B. 2002. Letter to Terry G. Jordan-Bychkov. Knoxville, TN, June 2nd.

Rehder, John B., John Morgan, and Joy L. Medford. 1979. "The Decline of Smokehouses in Grainger County, Tennessee." *West Georgia College Studies in the Social Sciences*. 18: 75–83.

Riedl, Norbert F., Donald B. Ball, and Anthony P. Cavender. 1976. *A Survey of Traditional Architecture and Related Material Folk Culture Patterns in the Normandy Reservoir, Coffee County, Tennessee*. Knoxville: University of Tennessee, Department of Anthropology.

Roark, Michael. 1990. "Cultural Areas of Missouri." *North American Culture*. 6: 13–24.

Roberts, Warren E. 1985. *Log Buildings of Southern Indiana*. Bloomington: Trickster Press.

Rose, Gregory S. 1985. "Hoosier Origins: The Nativity of Indiana's United States-Born Population in 1850." *Indiana Magazine of History*. 81: 201–232.

Rose, Gregory S. 1986. "Upland Southerners: The County Origins of Southern Migrants to Indiana by 1850." *Indiana Magazine of History*. 82: 242–263.

Sauer, Carl O. 1920. *The Geography of the Ozark Highland of Missouri*. Geographic Society of Chicago, Bulletin 7. Chicago: University of Chicago Press.

Schimmer, James R., and Allen G. Noble. 1984. "The Evolution of the Corn Crib." *Pioneer America Society Transactions.* 7: 21–33.

Schroeder, Walter A. 1982. *Presettlement Prairie of Missouri.* 2nd ed. Jefferson City: Conservation Commission of the State of Missouri.

Schultz, LeRoy G. 1983a. "West Virginia Cribs and Granaries." *Goldenseal.* 9(1): 47–54.

Schultz, LeRoy G. 1983b. "Log Barns of West Virginia." *Goldenseal.* 9(2): 40–45.

Scofield, Edna. 1936. "The Evolution and Development of Tennessee Houses." *Journal of the Tennessee Academy of Science.* 11: 229–240.

Sculle, Keith A., and Wayne H. Price. 1993. "The Traditional Barns of Hardin County, Illinois." *Material Culture.* 25(1): 1–27.

Semple, Ellen C. 1901. "The Anglo-Saxons of the Kentucky Mountains." *Geographical Journal.* 17: 588–623.

Sexton, Rocky. 1991. "Don't Let the Rain Fall on My Face: French Louisiana Gravehouses in an Anthropo-Geographical Context." *Material Culture.* 23(3): 31–46.

Shoemaker, Floyd C. 1955. "Missouri's Tennessee Heritage." *Missouri Historical Review.* 49: 127–142.

Shortridge, James R. 1977. *Kaw Valley Landscapes.* Lawrence, KS: Coronado Press.

Shortridge, James R. 1995. *Peopling the Plains: Who Settled Where in Frontier Kansas.* Lawrence: University Press of Kansas.

Smith, Gerald. 2002. Letter to Terry G. Jordan-Bychkov. Sewanee, TN, March 11th.

Smith, Joseph. 1854. *Old Redstone; or, Historical Sketches of Western Presbyterianism, its Early Ministers, its Perilous Times, and its First Records.* Philadelphia: Lippincott, Grambo & Co.

Smithwick, Noah. 1983. *The Evolution of a State, or Recollections of Old Texas Days.* Austin: University of Texas Press (originally published 1900).

Steinert, Wilhelm. 1999. *North America, Particularly Texas in the Year 1849: A Travel Account.* Dallas, TX: De Golyer Library and Clements Center for Southwest Studies at Southern Methodist University (originally published 1850).

Stewart-Abernathy, Leslie C. 1985. *Independent but not Isolated: The Archeology of a Late Nineteenth Century Ozark Farmstead.* Pine Bluff: Arkansas Archeology Survey.

Stuck, Goodloe R. 1971. "Log Houses in Northwest Louisiana." *Louisiana Studies.* 10: 225–237.

Sutherland, David. 1972. "Folk Housing in the Woodburn Quadrangle, Kentucky." *Pioneer America*. 4(2): 18–24.

Swanton, John R. 1928. "Social Organization and Social Usages of the Indians of the Creek Confederacy." *Annual Report of the Bureau of American Ethnology, 1924–1925*. 42: 23–472.

Swanton, John R. 1931. *Source Material for the Social and Ceremonial Life of the Choctaw Indians.* Washington, D.C.: Smithsonian Institution, Bureau of American Ethnology, Bulletin 103.

Swanton, John R. 1946. *The Indians of the Southeastern United States.* Washington, D.C.: Smithsonian Institution, Bureau of American Ethnology, Bulletin 137.

Tebbetts, Diane. 1978. "Traditional Houses of Independence County, Arkansas." *Pioneer America*. 10(1): 36–55.

Thomas, William R. 1930. *Life Among the Hills and Mountains of Kentucky*. Louisville: Standard Printing.

Thompson, S. H. 1910. *The Highlanders of the South.* New York: Eaton & Mains.

Torma, Carolyn, and Camille Wells. Ca. 1985. *Pulaski County Architectural and Historical Sites*. N.p.: Kentucky Heritage Council.

Transeau, Edgar N. 1935. "The Prairie Peninsula." *Ecology*. 16: 423–437.

Treat, Victor H. 1967. "Migration into Louisiana, 1834–1880." Ph.D. dissertation. University of Texas, Austin.

Tyner, James W., and Alice T. Timmons. 1970. *Our People and Where They Rest.* 2 volumes. Norman: University of Oklahoma American Indian Institute.

United States Department of Agriculture. 1939. *Farm Housing Survey*. Miscellaneous Publication No. 323. Washington D.C.: Government Printing Office.

Upton, Elsie. 1941. "The Austin Hill Folk." *Texas Folk-Lore Society Publication*. 17: 40–48.

Vance, Rupert B. 1932. *Human Geography of the South: A Study in Regional Resources and Human Adequacy*. Chapel Hill: University of North Carolina Press.

Veselka, Robert E. 2000. *The Courthouse Square in Texas.* Ed. Kenneth E. Foote. Austin: University of Texas Press.

Vlach, John M. 1972. "The Canada Homestead: A Saddlebag Log House in Monroe County, Indiana." *Pioneer America*. 4(2): 8–17.

Voegelin, Erminie Wheeler. 1944. *Mortuary Customs of the Shawnee and Other Eastern Tribes*. Indianapolis: Indiana Historical Society, Prehistory Research Series, II, no. 4.

Walz, Robert B. 1952. "Migration into Arkansas, 1833–1850." M.A. thesis. University of Texas, Austin.

Walz, Robert B. 1958. "Migration into Arkansas, 1834–1880." Ph.D. dissertation. University of Texas, Austin.

White, William W. 1948. "Migration into West Texas, 1845–1860." M.A. thesis. University of Texas, Austin.

Wigginton, Eliot (ed.). 1972. *The Foxfire Book.* Volume 1. Garden City: Anchor/Doubleday.

Wilhelm, Eugene J., Jr. 1965. "The Cultural Heritage of the Blue Ridge." *Mountain Life & Work.* 40(2): 16–20.

Wilhelm, Eugene J., Jr. 1966. "Animal Drives in the Southern Highlands." *Mountain Life & Work.* 42(2): 6–11.

Wilhelm, Eugene J., Jr. 1974. "The Mullein: Plant Piscicide of the Mountain Folk Culture." *Geographical Review.* 64: 235–252.

Wilhelm, Eugene J., Jr. 1978. "Folk Settlements in the Blue Ridge Mountains." *Appalachian Journal.* 5(2): 204–245.

Wilhelm, Eugene J., Jr. 1993. "Material Culture in the Blue Ridge Mountains." *Geoscience and Man.* 32: 197–256.

Wilhelm, Hubert G. H. 1972. "Southeastern Ohio as a Settlement Region: An Historical-Geographical Interpretation." *Pioneer America Society Proceedings.* 1: 95–123.

Williams, Michael Ann. 1987. "Rethinking the House: Interior Space and Social Change." *Appalachian Journal.* 14: 174–189.

Wilson, Eugene M. 1970. "The Single Pen Log House in the South." *Pioneer America.* 2(1): 21–28.

Wilson, Eugene M. 1975. *Alabama Folk Houses.* Montgomery: Alabama Historical Commission.

Withers, Robert S. 1950. "The Stake and Rider Fence." *Missouri Historical Review.* 44: 225–231.

Wright, Martin. 1950. "The Log Cabin in the South." M.A. thesis. Louisiana State University, Baton Rouge.

Wright, Martin. 1956. "Log Culture in Hill Louisiana." Ph.D. dissertation. Louisiana State University, Baton Rouge.

Wright, Martin. 1958. "The Antecedents of the Double-Pen House Type." *Annals of the Association of American Geographers.* 48: 109–117.

Zelinsky, Wilbur. 1951. "Where the South Begins: The Northern limit of the Cis-Appalachian South in Terms of Settlement Landscape." *Social Forces.* 30: 172–178.

Zelinsky, Wilbur. 1953a. "The Settlement Patterns of Georgia." Ph.D. dissertation. University of California, Berkeley.

Zelinsky, Wilbur. 1953b. "The Log House in Georgia." *Geographical Review*. 43: 173–193.

Zelinsky, Wilbur. 1977. "The Pennsylvania Town: An Overdue Geographical Account." *Geographical Review*. 67: 127–147.

Zelinsky, Wilbur. 1989. "Culture Areas." In David J. Cuff, William J. Young, Edward K. Muller, Wilbur Zelinsky, and Ronald F. Abler (eds.). *The Atlas of Pennsylvania*. Philadelphia: Temple University Press, 154.

Zelinsky, Wilbur. 1994. "Gathering Places for America's Dead: How Many, Where, and Why?" *Professional Geographer*. 46: 29–38.

(Illustrations are noted in *italics*.)

About the Author

Terry G. Jordan-Bychkov was born and reared in Dallas, Texas, in the Texas appendage of the Upland South. His maternal ancestral roots lie in the Carolina and Georgia Piedmont and in the southern Appalachians, with deeper roots in Celtic lands such as Ireland, Scotland, and Wales, as well as in England. Such people created the Upland South. His parents left the land and became city folks, but he recalls visits to his grandparental farm in East Texas, with its "dogtrot" house and other appurtances of upland southern life.

Choosing the life of an academician, he became a cultural geographer. Since 1982 he has been the Walter Prescott Webb Professor of History and Ideas in the department of geography at the University of Texas in Austin. He is the author or coauthor of more than a dozen scholarly books, among them *Texas Log Buildings* (Texas, 1978), *Texas Graveyards* (Texas, 1982), *American Log Buildings: An Old World Heritage* (North Carolina, 1985), with Matti Kaups, *The American Backwoods Frontier: An Ethnic and Ecological Interpretation* (Johns Hopkins, 1989), *North American Cattle Ranching Frontiers: Origins, Diffusion, and Differentiation* (New Mexico, 1993), with Bella Bychkova Jordan, *Siberian Village: Land and Life in the Sakha Republic* (Minnesota, 2001), and, with Alyson L. Greiner, *Anglo-Celtic Australia: Colonial Immigration and Cultural Regionalism* (Center for American Places, 2002).

Jordan-Bychkov has won numerous awards for his research and has served in the honorific position of president of the Association of American Geographers. He is a fellow of the Texas State Historical Association.

The Center for American Places is a tax-exempt 501(c)(3) nonprofit organization, founded in 1990, whose educational mission is to enhance the public's understanding of, and appreciation for, the natural and built environment. It is guided by the belief that books provide an indispensible foundation for comprehending—and caring for—the places where we live, work, and explore. Books live. Books endure. Books make a difference. Books are gifts to civilization.

With offices in New Mexico and Virginia, Center editors bring to publication 20–25 books per year under the Center's own imprint or in association with publishing partners. The Center is also engaged in numerous other educational programs that emphasize the interpretation of *place* through art, literature, scholarship, exhibitions, lectures, curriculum development, and field research. The Center's Cotton Mather Library in Arthur, Nebraska, its Martha A. Strawn Photographic Library in Davidson, North Carolina, and a ten-acre reserve along the Santa Fe River in Florida are available as retreats upon request.

The Center strives every day to make a difference through books, research, and education. For more information, please send inquiries to P. O. Box 23225, Santa Fe, NM 87502, U.S.A. or visit the Center's Web site (*www.americanplaces.org*).

About the Book:
The text for *The Upland South: The Making of an American Folk Region and Landscape* was set in Bembo. The paper is 80# Domtar Plainfield.

For The Center for American Places:

George F. Thompson, president and publisher

Randall B. Jones, associate editorial director

Purna Makaram, manuscript editor

Martin F. Jones, indexer

David Skolkin, designer and typesetter

Dave Keck, of Global Ink, Inc., production coordinator